The Dolphins Are Back

The Dolphins Are Back

A Successful Quality Model
for Healing the Environment

Phillip M. Scanlan

Productivity Press
Portland, Oregon

Additional copies of this book are available from the publisher. Discounts are available for multiple copies through the Sales Department (800-394-6868). Address all other inquiries to:

Productivity Press
P.O. Box 13390
Portland, OR 97213-0390
United States of America
Telephone: 503-235-0600
Telefax: 503-235-0909
E-mail: service@ppress.com

Text design by Bill Stanton
Cover design by Christopher Hanis
Art composition by Gordon Ekdahl, Fineline Graphics
Page composition by William H. Brunson Typography Services

Photo credits: *text*: p. 3, © by Herbert Segars; p. 118, Robert Schoelkopf; p. 200, courtesy of the White House Press Office. All photos printed with permission. *jacket:* cover photo by Steve Bloom/Masterfile; author photo by Gail Hannagan

Printed and bound by Malloy Lithographing in the United States of America
Text is printed on 85% recycled paper.

Library of Congress Cataloging-in-Publication Data

Scanlan, Phillip M.
    The dolphins are back : a quality model for successfully forming business, government, and citizen participation partnerships to create environmental improvement programs / Phillip M. Scanlan.
      p. cm.
    Includes bibliographical references and index.
    ISBN 1-56327-183-4 (hardcover)
    1. Marine pollution—New Jersey—Atlantic Coast.   I. Title.
GC1212.N5S33   1988
363.739′45′09749—dc21   97-48648
                                          CIP

Royalties from the sales of this book to AT&T organizations will be paid by the publisher to the Center for Marine Conservation (CMC). The CMC is a nonprofit, environmental group that is focused on cleaning up beaches and oceans on a worldwide basis. The address is: Center for Marine Conservation, 1725 De Sales Street, NW, Washington, DC 20036.

02 01 00 99 98   10 9 8 7 6 5 4 3 2 1

Dedication

Quality improvement is dedicated to making things better for the next year. Environment improvement is dedicated to making things better for the next generation.

This book is dedicated to my two grandsons—Trevor, five; and Travis, three.

Contents

Publisher's Message

Phil Scanlan's book, *The Dolphins Are Back: A Successful Quality Model for Healing the Environment*, is a direct response to a disappointing summer vacation he had with his eleven-year-old son in 1988—when a New Jersey shore beach was closed due to pollution. Phil, the Vice President of Quality at AT&T, subsequently learned that from June 1987 until March 1988 natural and human-made toxins and bacteria caused the deaths of some 2,500 to 3,000 bottle-nosed dolphins along the Atlantic coast from New Jersey to Florida. The future of the New Jersey shore looked bleak. Beach closings due to pollution were at an all-time high and there was no one person to hold responsible or anyone who had the authority to initiate a solution. The people affected by and involved with the pollution operated in an atmosphere of mistrust, blaming big business, the federal government, New Jersey State, New York City, tourists, motor boats, etc. Meanwhile the government's approach of using fines wasn't improving the quality of the water. People who were aware of the extent of the problem had begun to accept that achieving clean water was no longer a possibility. As late as 1995 the U.S. House of Representatives was pushing forward legislation that would have circumvented much of the Clean Water Act of 1972.

When Phil Scanlan had the opportunity in 1989 to volunteer for a newly formed group, Quality New Jersey (QNJ), to improve the quality of New Jersey he brought something unique to the New Jersey shore problem—the AT&T total quality methodology. As quality vice president at AT&T, where he has been employed since 1966, Phil Scanlan has been instrumental in developing an AT&T total quality approach and the highly acclaimed AT&T Quality Library. He developed the AT&T Core Quality Curriculum of more than a dozen quality courses approved for college credits. It made sense for QNJ to use his expertise to apply total quality methods to the problems of the New Jersey shore. Businesses had been successfully implementing continuous improvement programs for years and since 1987 the U.S. Commerce Department had been recognizing them for such efforts. There was no reason why these total quality methods shouldn't also work on a larger complex scale like cleaning up the environment.

The Dolphins Are Back is a story about the kind of dedication it takes to effect positive change. It is also the story of what happens when caring citizens like Phil Scanlan put their mind to achieving something that everyone said was impossible: using quality principles to bring business and government together to clean up one of the worst shores in the United States. Ultimately, *The Dolphins Are Back* is an important lesson of how people, when provided with the proper methodology, can have a great effect on continuously improving a management system's ability to tackle a problem or achieve a result. In addition the book underscores the importance of recognizing the interdependency among business, government, nonprofits, and citizens and the persistent effort needed to communicate and work together when dealing with large, complex problems such as the environment. As Phil discovered early on in leading QNJ's citizen-business-government team, people already care. They *are* interested and want improvement—they just need an effective improvement process model they can understand and apply so they can move from the "blame" environment to a "solution" environment.

In *The Dolphins Are Back*, Phil Scanlan tells two success stories: the development and application of the quality methodology used at AT&T and the use of this same methodology to clean up the

New Jersey shore. It is a story about what it takes to create a culture of continuous improvement in a business-government-citizen partnership. Where many quality improvement books are skill-based and include a step-by-step methodology for the application of a particular quality tool or method, this book uses a few large, complex, real-life examples to provide an awareness of the overall quality approach as well as discusses the significant deployment issues that were encountered. As Phil says, "quality improvement is 20 percent approach and 80 percent deployment."

In *The Dolphins Are Back* you will follow the collaborative process of quality improvement of AT&T and the New Jersey shore over a nine-year period. You'll follow the many challenges of identifying the root causes and implementing quality control amidst the ever-changing landscape of local, state, and national politics. You will learn how QNJ used the AT&T quality tools to establish process management capabilities; how they set out to define the pollution process; how they identified the customer requirements and determined who was responsible for the pollution. You will learn how they defined and established process measures for the environment and assessed conformance to these requirements so they could investigate the process to identify improvement opportunities. Along the way you'll encounter QNJ's many challenges and surprises. For instance, there was no one major reason or entity responsible for the pollution. And QNJ's Pareto analysis revealed that over 95 percent of the beach closings in 1988 was not caused by the oil spills or New York's debris that the news coverage and politicians had been focusing on, but by New Jersey's inadequately maintained sewage and storm drain system. In other words, the "process owners" were the 90 municipalities along the 127-mile New Jersey ocean coastline. Every time it rained the incidences of pollution and beach closings increased.

Throughout the book you'll follow the story of QNJ facilitating, supporting, and motivating the 90 New Jersey shore municipal leaders and the five New Jersey shore county leaders to work in a culture of continuous improvement. You will see how they rewarded shore communities who made improvements by developing the annual QNJ Shore Quality Award which covers two of the seven Baldrige category areas: prevention processes and shore water quality results. And all the while you are learning how they were applying quality methods to the New Jersey shore you'll be following the quality successes at AT&T—three Baldrige awards, a Deming prize, two Shingo prizes and several state quality awards.

Phil's and QNJ's success in using the AT&T quality approach and coordinating the efforts of four levels of government provide businesses, governments, nonprofits, environmental groups, and shore communities all across the world with a workable collaborative quality model. We hope the reader takes time to learn how the dolphins came back, for it is because of the dolphins that attention was focused on the pollution along the Atlantic Coast. The New Jersey shore is now one of the cleanest shores in the United States, providing an empirical example *that we can improve the quality of our water.*

We would like to thank Phil Scanlan for bringing this unique book to Productivity Press. This is one of the first books we've published on total quality control that steps out of the manufacturing arena into a business-government application. We also wish to thank all those who participated in shaping this manuscript and bringing it to bound book: Diane Asay, editor in chief; Jessica Letteney, prepress manager; Gary Peurasaari, development editor; Susan Swanson, production editor; Chris Hanis, cover designer; Dave Lindstedt and Jane Loftus, copyeditors; Sheryl Rose, proofreader; William H. Brunson of Typography Services, typesetter; Gordon Ekdahl, art designer; and Bill Stanton, text designer.

Steven Ott
President and Publisher

Preface

The purpose of this book is to share lessons learned from applying a total quality approach, in both a business and a government application, to achieve a culture of continuous improvement. People who are interested in how to achieve a culture of continuous improvement in business, government, or nonprofit organizations could find this book both useful and interesting.

This book is based on my nine years of experience as the quality vice president at AT&T and eight years of experience leading a citizen–business–government team that focused on cleaning up New Jersey's shore water quality.

This citizen–business–government team, part of a volunteer nonprofit organization called Quality New Jersey (QNJ), has applied the AT&T quality approach to the problem of cleaning up the worst ocean water-quality problem in the nation. With the support of QNJ, the state of New Jersey and New Jersey shore counties, the New Jersey shore municipalities achieved the most improved ocean water quality in the nation.

In striving to educate AT&T employees on how to apply our AT&T quality approach, the most consistently effective example was the QNJ ocean water improvement project. This example has been an effective model because people care about the shore and are interested in its improvement. Consequently, they want to understand the quality improvement process that was used and then try to apply it to other situations. With many improvement projects there is often a feeling of hopelessness because of the size or complexity of the problem, or because the problems have been around for so long. It is easy to believe that a problem is too big, too complex, or too pervasive to ever be solved. But the Jersey shore example, which involved coordinating the efforts of four levels of government, demonstrates how a quality approach can be used to solve even a long-standing, large, and complex problem like cleaning up ocean water quality.

The improvement in New Jersey's shore water quality demonstrates how AT&T's quality approach can be used to achieve a culture of continuous improvement in business or government for customers, employees, investors, or taxpayers. In this book we will also look at the application of a quality approach within AT&T—specifically, the continuous improvement of communications network quality.

I have found that a significant obstacle to implementing quality improvement is the difficulty of discussing problems and their root causes without sounding critical. These root causes are typically embedded in the current system or process that people are using, and it is usually the system or process itself that must be improved, not the people who are using it. In order to improve our quality at the rate necessary to compete in the world today, we must learn to discuss problems and their root causes openly. The intent of this book is not to criticize people or organizations, but to demonstrate how to apply a quality approach to eliminate root causes of system and process problems. Data must be used not to blame, but to identify causes that can be eliminated through process improvements.

The application of a total quality approach in a public institution such as government is similar (with important differences that will be addressed) to the application in a private business institution. In fact, most of our businesses are "public institutions," because they are owned by the public and subject to public ownership requirements. Our business CEOs face competition in the marketplace, and they may be replaced by the company's public owners if they do not produce the desired results. In 1993 alone, the CEOs of American Express, IBM, Westinghouse, Apple Computer, Eli Lilly, Eastman Kodak, Scott Paper, and Borden were all replaced, in no small measure due to pressure from institutional investors who represent public ownership and who were dissatisfied with the performance of these businesses. Middle managers and operations workers face possible job loss if these businesses cannot compete in the marketplace. Jobs may be transferred elsewhere or outsourced to a more focused and efficient firm in the area, or in another state or country.

Our government executives and legislators face competition at each election. They, too, must produce the results the public wants or risk being replaced. With pressure on the government to balance budgets and produce results, government workers face outsourcing as well. Businesses may leave a state or country that is not providing excellent services at a reasonable cost, thereby depriving the state or country of a valuable source of jobs and revenue.

Quality methods can be applied by managers in both private and public organizations to achieve a culture of continuous improvement. In this book I share my experience applying a total quality approach to achieve excellence in both private and public organizations. These insights gained have been summarized at the end of each chapter as Lessons Learned. Though you could simply read the Lessons Learned and gain an overview of our accomplishments, by reading the entire book you will understand the lessons more fully, and you will come to appreciate the critical importance of a commitment to deployment, even in the face of many barriers.

There are no guarantees that by using a quality approach for continuous improvement we will achieve and sustain our goal of excellence. Several major factors determine whether the goal of excellence will be achieved and maintained. Key influences include external factors, such as changes in regulations (the rules of the game set by government), competitive changes, and changes in technology and customer requirements. Internal factors, such as corporate strategies, acquisitions, reorganizations, leadership changes, and the management system itself, also come into play.

In government applications, the external factors are such issues as world stability and trading agreements (the rules of the game); world, national, and local economies (which are driven by business); competition for global business investments; in various market sectors, such as agriculture, manufacturing, services, and information. The internal factors affecting government include political strategies, government policies, reorganizations, leadership, and management systems.

Nonprofit organizations are affected by external factors such as the economy; government, business, and citizen support; and environmental and social problems. Internal factors for nonprofits include strategies, reorganizations, leadership changes, volunteer help issues, and the organization's management system.

Whether in business, government, or a nonprofit organization, we all have common issues to face in achieving a culture of continuous improvement. We have little control over most external factors and little control over the major internal factors that influence the achievement of excellence. However, most of us can have an effect on the quality of a management system by using a quality approach to continuously improve it. The better our management system is, the better the chances are of achieving excellence; if we continuously improve our management system, we improve our chances of surviving through the tough times and going on to achieve another day—as most excellent companies have done. Life in business is not without its ups and downs. But a good management system, which is obtained through a quality approach to continuous improvement, can help to ease the up and down ride and keep us more often on the up side than on the down side.

In addition, we must recognize that there is an interdependency between business, government, and nonprofits. The external factors of one are often the internal factors of another. Therefore, as we improve our organization's internal factors, we also improve the external factors for someone else. Improvement in one area does not necessarily come at the expense of another area; it could be a benefit for the other, as well.

Many books on quality improvement are skill-based and include a step-by-step methodology for the application of a particular quality tool or method. These books often are illustrated by simple, academic examples that make clear the proper use of the tools, and the objective is to teach the proper use of the tool or method. In contrast, this book uses a few large, complex, real-life examples to provide an awareness of the overall quality approach and to discuss the significant deployment issues that were encountered. Quality improvement is 20 percent approach and 80 percent deployment. This book explains what it takes to deploy and sustain a quality approach to achieve a culture of continuous improvement.

Acknowledgments

I would like to acknowledge the many people who contributed to the quality improvement efforts described in this book and/or to the writing of this book:

The People in These Organizations:

AT&T
Quality New Jersey (QNJ)
Malcolm Baldrige National Quality Award
Northeast Regional Environmental Protection Agency (EPA)
New Jersey Department of Commerce
New Jersey Department of Environmental Protection (DEP)
New Jersey shore counties and municipalities
New Jersey shore watershed planning commissions
Rutgers University
Tufts University

The People in These Environmental Associations:

Association of New Jersey Environmental
    Commissions (ANJEC)
Center for Marine Conservation (CMC)
Natural Resources Defense Council (NRDC)
New Jersey Future

The People in These Teams:

AT&T Quality Directors and Quality Community
AT&T Quality Office Team Members
QNJ Environment Team Members

Individuals:

QNJ Environment Team:	Dave Rosenblatt, Peter Brandt, Laurie Groves, Judy Soltis, Patricia Walsh, Don Haberstroh, and Rae Hoopes
QNJ Organization:	Dom Facchini, Judy Finman, Paul DeBaylo, Ed Shecter, Brent Ruben, Lawrence Codey, Vince Franco, and Rich Serfass
NJ State Government:	Gualberto Medina, Robert Shinn, and Christine Whitman
AT&T:	Bob Allen, Frank Ianna, Ken Bertaccini, and Nick Finamore
AT&T Quality Office:	Dale Myers, Susan Annitto, Lil Findlay, Frenck Waage, and Karla Monson
Friends:	Jack Radochia, Charles Selden, Peggy Dellinger, and Bill Ebbihara
Family:	Jane, Scott, Maury, Michelle, and John

CHAPTER ONE

This Beach Is Closed
(1987/1988)

My son John and I were driving down Main Street on Long Beach Island during the latter part of the summer of 1988, trying to find a beach house for a week at the shore. We had our requirements. It had to be right on the shore with a view of the ocean. It had to be within walking distance of miniature golf and the movies—John was eleven. It had to sleep about 8 to 10 people, allowing for my family of six plus a few friends of the kids. And it had to cost no more than $500 per week. We were cruising the strip, checking each Realtor for a cancellation that we could pick up for half price.

After spending the hottest part of the day with Realtors, we found a place that met our requirements. Our last-minute vacation planning was a success.

When we got home that evening, I expected enthusiastic support, but our great buy did not meet everyone's requirements. No dishwasher, no cable TV or VCR, no phone, no air conditioner! But, it was right on the beach and half price!

This was our first trip to the New Jersey shore for summer vacation. My parents had a cottage on a lake in New Hampshire that had been *the* summer vacation spot for my whole life. While vacationing in New Hampshire, my kids got to spend time with their Nana and Papa, along with the use of the rowboat and sailboat that I had bought when I was a kid. The Jersey shore would have to be good to replace Nana's and Papa's lake house.

When we arrived at the shore on a Saturday, it was raining. It rained the first two days of our week at the shore—not a great start when you have four kids ages 11 to 19. The third day, Monday, was sunny and everyone enjoyed the beach and the water, although the water wasn't very clear.

On the fourth day, a sign was posted on the beach: "This beach is closed due to pollution." The beach stayed closed the rest of our week at the shore.

My vacation planning wisdom was now being questioned by my disappointed teenagers. We were only able to go in the water one day. Not even Chevy Chase would have dragged his family on this "Summer Vacation."

The next week's news coverage of the polluted Jersey shore showed oil slicks, syringes, garbage, and tons of things floating off-shore or washed up on the beach.

WHY DID THE DOLPHINS DIE?

From early June 1987 until March 1988, unprecedented numbers of dead and dying bottle-nosed dolphins washed ashore along the Atlantic coast from New Jersey to Florida (Photo 1-1).

In mid-June 1987, Mr. Robert Schoelkopf, director of the non-profit Marine Mammal Stranding Center in Brigantine, New Jersey, became suspicious. Mr. Schoelkopf alerted Dr. James Mead of the Smithsonian Institution, who oversees the world's largest collection of marine mammals. By the end of July, Dr. Mead had information on 30 dead dolphins. Mr. Schoelkopf said, "Normally, during the course of a summer season, you see three or four dolphins wash up dead, and for each one accounted for, three go unnoticed because sharks eat them."[1]

In late July, Dr. Mead persuaded environmental agencies to get help from Dr. Joseph Geraci, a veterinary pathologist at the University of Guelph in Ontario, Canada. A few days later, Dr. Geraci set up camp at the Virginia Marine Science Museum in Virginia Beach, and established a network of 34 experts at 19 institutions.[2]

Photo 1-1. Dead Dolphin on the Beach

By March 1988, 740 dolphins had washed ashore and died. If only one out of four dolphins that died reached the shore, then the total dolphin death count may have been in the range of 2500 to 3000. It has been estimated that 50 percent or more of the migratory coastal dolphin stock between Florida and New Jersey died during this period.[3] The two most likely causes for death were poison or an infectious disease.

Natural Toxins and Bacteria

In April 1989, Dr. Geraci completed his final report on his extensive investigation: Forty-two pages filled with descriptions of the sad and painful death that thousands of dolphins faced as they struggled to fight off the overpowering effects of a combination of poison and infectious diseases. The happy, playful, smiling dolphins that we know and love from the aquarium and movies were not happy during the summer of 1987. The following is a summary from the investigative report:

The dolphins that came ashore in August and early September had a range of skin lesions. Small blisters and pock-like craters were common over the head region. A viral infection was suspected and a papovavirus was detected in 4 of 12 dolphins examined.

A second type of skin lesion killed large areas of skin, exposing underlying, intensely reddened, inner skin. The epidermis could be peeled back as easily as cellophane. These lesions were caused by bacteria, fungi or protozoa. This condition was one manifestation of a systemic bacterial invasion that seemed to be the ultimate cause of death for many of the dolphins during the hot summer months. Dolphins with this condition were found only during the period between June 1987 and March 1988, while dolphins with the viral lesions continued to be found after this period.

Blood vessels had been injured by the bacteria. The vessel walls became fragile and were unable to contain blood. Plasma leaked into tissue spaces. Affected organs died from a combined effect of impaired circulation and bacterial toxins. In some cases, a protracted illness resulted in pneumonia, cerebral hemorrhage, secondary invasion by fungi organisms, vascular collapse, or shock. Chronic lesions were found in several dolphin deaths. These lesions were in the lungs, liver, pancreas, and heart, and were characterized by fibrosis.

Some of the dolphin liver samples contained brevetoxin (PbTx), a poisonous natural toxin produced by a red-tide organism. The red tide is caused by a one-cell dinoflagellate plant called *Gymnodinium breve*. This potent brevetoxin was found in 8 of the 17 (47 percent) dolphins examined. Most of the dolphins did not die from the amount of natural toxin they consumed, but from diseases affecting the liver and lungs and from secondary infections associated with immune suppression. It is possible that "sublethal" exposure to the natural red-tide toxin (PbTx) affected the immune system and precipitated the train of events leading to some or all of these chronic changes. One explanation of how the dolphins ate this natural toxin is that they may have encountered it along their northerly migration route from Florida to New Jersey.

While the red-tide toxin was viewed as one of the causes of the deaths of the dolphins in 1987, it is unlikely that this was the first time the dolphins were exposed to this toxin, since it blooms regularly. Therefore, it alone could not account for this major dolphin kill. The deaths of half the North Atlantic dolphins were felt to be the result of a combination of natural toxins and bacteria simultaneously present in the East Coast waters.

Virus and Bacteria

In 1994, another study was published by Thomas Lipscomb, Department of Veterinary Pathology, Armed Forces Institute of Pathology, Washington, D.C. along with three members of the Veterinary Sciences Division, Stormont, Belfast, Ireland.[4] At the direction of the U.S. National Marine Fisheries Service, tissue from 80 bottle-nosed dolphins collected by Dr. Geraci during the 1987 to 1988 incident were transferred to the Armed Forces Institute of Pathology for independent investigation and permanent storage. This study examined the dolphins for viral infections, using a different technique than the one used by Dr. Geraci, and found that about half of the dolphins had a morbilliviral infection that was not found in the earlier study by Dr. Geraci. Morbillivirus causes a suppression of the immune system, which made the dolphins more vulnerable to bacterial and fungal infections. Between 1988 and 1994, other episodes involving major dolphin deaths occurred in the Mediterranean, the Pacific, and the Gulf of Mexico. Therefore, Dr. Lipscomb concluded that the morbillivirus and the bacterial infections were likely contributing causes of the dolphin deaths in the June 1987 to March 1988 period. It is believed the dolphins that survived this illness developed an antibody that will protect them against a reoccurrence of this particular virus.

Both the 1989 and 1994 studies agreed that the bacterial infections caused the deaths, but each study had a different conclusion for the cause of the immune system suppression that led to the deaths from the bacterial infection. The 1994 study concluded that the immune system suppression was caused by a virus that was undetected in the 1989 study. The 1989 study concluded that the primary cause of the immune system suppression was the natural red-tide toxin. In all likelihood, both the virus and the toxin played a major role.

Man-made Toxins and Bacteria

In a still later study researchers from the Skidaway Institute of Oceanography in Savannah, Georgia; Sea World of Orlando, Florida; Michigan State University; and Ehime University of Japan, found significant doses of tributyl-tin (TBT) and related tributyl compounds in the livers, kidneys and muscles of the stranded dolphins from the June 1987 to March 1988 deaths.[5] Tissue samples were tested from the dolphins that had been kept in storage at the Armed Forces Institute of Pathology.

Tributyl compounds were added to marine paints from the 1960s until 1988 to prevent attachment of barnacles and slime. Their use on boats smaller than 82 feet was banned in 1988 after they were shown to be toxic to oysters and other marine life. These compounds suppress mammal immune systems, which in turn makes them susceptible to infections, like pneumonia. Based on analysis of tissues taken from stranded wild dolphins who died in the June 1987 to March 1988 period, it was found that they had 50 to 100 times the amount of butylins compared to dolphins that were in aquariums when they died. Unlike PCBs, which accumulate in dolphin and whale blubber, the butylins were found in vital organs.

It is probable that the butylin compounds, in addition to PCBs, have contributed to the immune system suppression that led to the dolphin deaths off the Atlantic coast, says the research team led by Kururntha Kannan of Michigan State University.

After three separate studies, three causes were found for the dolphin immune system suppression (a natural red-tide toxin, a virus, and a butylin chemical in marine paint), each of which contributed to the deaths of the dolphins whose immune systems could not fight off subsequent infections.

WHY DID WE HAVE THE TOXINS AND DISEASE?

In 1997, *And the Waters Turned to Blood,* by Rodney Barker, was published describing the work of JoAnn Burkholder, an aquatic ecologist at North Carolina State University.[6] Dr. Burkholder studied river algae with the help of Howard Glasgow and found an organism that was associated with 30 percent to 50 percent of the fish kills in North Carolina waters between 1991 and 1993.

From thousands of dinoflagellates (two-tailed organisms), they found one, which they named *Pfiesteria piscicida,* that was producing a toxin that could kill fish caught in a heavy growth of this organism. While the red tide was a single-cell dinoflagellate "plant" that produced a toxin, this *Pfiesteria piscicida* organism was a single-cell dinoflagellate "animal" that spent most of its life in a dormant state in a river, or ocean, and feeds itself by stealing chloroplasts from plants. When a fish swims by, the *Pfiesteria* comes out of its dormant state, releases a toxic poison into the water, and propels itself, using its two tails, to eat the fish. As the fish dies, the organism burrows into the fish and, while feasting on the fish, simultaneously reproduces thousands of new dinoflagellates.

In 1995, this organism killed millions of fish and forced the closing of the Neuse River in North Carolina to recreation. It was also found that this organism would attack mammals, such as dolphins.

Dr. Burkholder found that the growth of this organism was caused by excessive nutrients that were overloading and polluting the water supply. The nutrients were from hog farm waste, fertilizers, a phosphorus mine along the river, and poorly maintained sewage treatment plants. In essence, the growth of the organism that produced a natural toxin was caused by our "fertilization" of the river and ocean with polluted runoff and sewage.

In 1985, there were 22 identified species of known toxic dinoflagellates. In 1997, there were 59, an increase in identified toxic killers of more than 150 percent in 12 years. Increased water pollution throughout the world caused increasingly large concentrations of these toxic killers, resulting in numerous fish and mammal kills, as well as human health problems. These toxic poisons can be transmitted to humans through contact in the water or through the air as sea breezes spread the toxins.

In late 1996, another study, using yet another new testing procedure, showed that a record number of manatee deaths in South Florida was caused by pneumonia that had developed after the mammals were weakened by the red-tide toxins that had accumulated in their lungs. The toxin caused the manatees to be unable to resist the bacterial infection that caused the pneumonia.

Created by the pollution of our waters, natural toxins were on the increase in number and in frequency of occurrence. The worse the environment that we create, the tougher the life forms become in order to survive. Whether that involves people in a drug and crime ridden inner city, soldiers in a war zone or organisms in polluted waters, the tough survive and the friendly die when the environment gets too rough. Life forms must adapt quickly to the environment in order to survive.

It was the combination of the immune system suppression and the subsequent bacteria-caused illnesses that resulted in the dolphin deaths.

Over time, many dolphins may have eaten enough butylin from the paint used on boats to weaken their immune systems. On their way north during the summer of 1987 they may also have eaten some of the natural toxins from the Florida toxic dinoflagellate single-cell plants, or been attacked by the North Carolina toxic dinoflagellate single-cell animals—toxic organism growth that was caused by polluted runoff water. Upon arriving in New Jersey, they may have encountered polluted waters containing both viruses and bacteria. The combination of these factors was apparently too much for about half the dolphins to fight off.

The causes of the immune suppression may never be proven. Perhaps it was the red-tide toxin from Florida, perhaps the *Pfiesteria* toxin from North Carolina, perhaps the morbilliviral virus, perhaps the butylin from the boat paint, or perhaps a combination of these four and other causes. However, all agree that the bacteria diseases are what killed the dolphins once they were weakened by a toxin or a virus.

Man-made pollution was the root cause of three of the four causes of the dolphin deaths: virus, bacteria and natural toxin growth. The fourth cause was a man-made toxin.

WHAT TO DO ABOUT IT?

At any time in the future, the dolphins' immune system could again be weakened by a natural toxin, a man-made toxin or by another virus, thereby making them vulnerable to attack by bacteria in our waters. One thing we could do is reduce the amount of pollution in our river and ocean waters that would reduce the natural toxins, the viruses, and the bacteria in our rivers and oceans.

The beaches are closed by the Health Department when a bacteria presence is measured above a certain count that could likely cause illness to humans. The measure used by the Health Department is the amount of fecal coliform found in the water. Fecal coliform is the bacteria found in human waste, or untreated sewage. The presence of this bacteria also indicates the likely presence of viruses.

My kids couldn't swim while on vacation because the water had a bacteria count that indicated it was unsafe for them. It was unfortunate for them, and me, that they couldn't go swimming, but fortunate for all that the Health Department protected them from potential disease in the water. Unfortunately, swimming isn't closed to the dolphins when we pollute the ocean. They get ill. They die.

The New Jersey shore pollution reached crisis levels and New Jersey citizens were angry about it. The blame for New Jersey's problems was being laid on business, waste disposal companies, and New York City. *They* were not doing a good job. *They* were destroying our New Jersey shore. Beach trash and litter were on the TV news and in newspaper stories throughout the summer of 1988. However, trash, oil, and other visible problems were not the cause of the dolphin deaths or the majority of the New Jersey beach closings. Less-visible bacteria, viruses, and toxins were the cause of the dolphin deaths and the cause of beach closings as a result of pollution of the Jersey shore waters.

The chemical toxin butylin, used in boat paint, has been banned since 1988 for small coastal boats, although the effect of this toxin will be present in our ocean, and in our dolphins, for years to come.

Scientists believe that the North Atlantic dolphins that survived the morbilliviral infection have developed antibodies that will protect them against a reoccurrence of that particular virus.

Viruses and bacteria, and even "natural" toxins, are the result of the heavily polluted waters that result from using the ocean as a dump for our sewage, sludge and polluted storm-water runoff. Sludge is the solid matter that remains after a sewage treatment plant has separated pollutants from clean water. Sludge contains concentrated amounts of pollutants.

One of the four causes of the dolphin deaths was addressed in 1988, by eliminating the man-made toxin in the boat paint. The remaining causes have pollution of the ocean water as a root cause. To address the root cause of the dolphins' deaths, we needed to address the pollution problem.

These toxins, viruses, and bacterial diseases combined to kill off half the North Atlantic bottle-nosed dolphins between 1987 and 1988. To avoid a repeat of massive numbers of dolphin deaths, we need to reduce the massive pollution of our shore waters. We need to stop using our coastal ocean waters as a dump for raw sewage, sludge, and polluted storm-water runoff that is loaded with viruses and bacteria.

In researching the reasons behind the dolphin deaths, I came across a poem written by Katharine Lubiejewski that describes the problem, and what we need to do about it, in a way that the scientific analytical studies do not capture. Katharine grew up in Lambertville, New Jersey, and always vacationed in Atlantic City. She told me that she was saddened by the plight of the dolphins during the year they suffered such anguish and pain. She respected dolphins as highly intelligent animals that are family oriented, and she imagined how they were trying to reason out their problem while trying to stay nearby to help their afflicted family members. She felt their helplessness, their pain, their confusion.

Save the Dolphins—Save the Sea
By Katharine Lubiejewski

As I stood there on the jetty,
Gazing far out to sea,
I watched the object drifting,
The tides bringing it to me.

As it moved closer to the shore,
I walked down onto the beach.
I moved father down the sand,
Until the object I had reached.

Then the tears welled in my eyes,
As I looked down where it was lying.
A splendid bottle-nose dolphin,
And I knew he was dying.

As I stood there at the water's edge,
Just him, and me, alone,

I'd never felt such helplessness,
Nor such guilt for what we've done.

For dolphins have for centuries
Loved to play 'longside our ships.
And guide our hapless sailors
Home from many ill-fated trips.

And how have we repaid them?
We've dumped sludge into their home.
We've also added chemicals,
Which kill them where they roam.

This truly special mammal,
Who has always been our friend,
Has always thrilled to see us,
Now we've brought him to an end.

I cannot hold my head up,
or feel proud on any day,
When I know we've killed the oceans
Where the dolphins used to play.

And I cannot help but wonder
About our polluted sea ...
If it can kill a nine-foot dolphin,
What can it do to me?

THE BEGINNING OF MY FOCUS ON QUALITY

In April 1988, I transferred from an AT&T business unit to a new position in the AT&T quality office. That same month, a new chairman, Bob Allen, took over at AT&T. Over the next several years, we established a partnership on our quality approach to achieving a culture of continuous improvement at AT&T.

Bob Allen provided the leadership and motivation for our improvement efforts, while my job was to provide the measures, education, and recognition support for our quality improvement effort. This involved enabling each of our associates to find the root causes of problems and to make improvements to prevent reoccurrence of problems. Fixing problems is merely repair, preventing problems is quality. Repairs are costly and problems leave customers unsatisfied. Preventing problems, by eliminating root causes, eliminates the cost of repair and increases customer satisfaction by reducing the number of problems. A quality approach results in higher sales at lower costs, producing satisfied customers, higher market share, and higher profits.

As the summer of 1988 closed, I wondered if our AT&T quality approach could also be used to improve the New Jersey ocean water quality. Could a business quality approach be used in government to prevent community problems? Could my kids go to the Jersey shore again on vacation and swim in clean water? Could we prevent pollution and another major dolphin kill?

Quality Arkansas—Sharing Quality Approaches

In the fall of 1988, I was invited to give the keynote talk at a quality conference in Little Rock, Arkansas. AT&T had a factory in Little Rock, and some of our AT&T people were volunteers supporting the annual Quality Arkansas Conference. Sharing AT&T's quality experiences with others was part of my job.

AT&T had much to share from its early work on quality pioneered by recognized AT&T quality leaders such as Shewhart, Deming and Juran. My keynote talk was not memorable, but the luncheon speaker at the Quality Arkansas Conference was— Governor Bill Clinton.

Governor Clinton joined the speakers at the head table at lunch and encouraged the Arkansas business people at the conference to continue their sharing of quality approaches, as a way to save jobs and improve the Arkansas economy.

During lunch, I was seated next to Governor Clinton. He demonstrated a knowledge and commitment to total quality management that surpassed that of many business executives at the time. He had been the luncheon speaker at this annual conference since it was started in the mid-1980s, and never missed one. The visible role of government leadership in encouraging the sharing of best practices and the continuous improvement of business is an important and underestimated factor in motivating continuous improvement.

Quality New Jersey (QNJ)—More Than Sharing

In early 1989, I got a call to attend a meeting with a few people who were considering starting a state-wide quality association in New Jersey that sounded similar to the Quality Arkansas group. Stan Marash and Ed Shecter, both New Jersey quality consultants, were looking for major New Jersey corporate involvement. In 1989, AT&T was the largest private business in New Jersey with 50,000 of our 300,000 people located there. The next largest New Jersey corporation was about half our size. I thought that the state with the largest concentration of AT&T employees deserved at least the same support we provided in Arkansas, where AT&T had only a small factory.

My stay at the New Jersey shore during the summer of 1988, and my visit to Arkansas during the fall of 1988, had a significant impact on my decision to help start the Quality New Jersey (QNJ) nonprofit organization in 1989.

A handful of people attended the first meeting and decided that the idea of sharing quality approaches in the state was useful, but that more than sharing could be done. A second meeting was planned around the idea of creating focus groups that would work to improve the quality of both work and life in New Jersey. Each focus group would focus on a selected sector of the New Jersey economy, such as the manufacturing sector or the services sector. In addition, Quality New Jersey would also have focus groups to help apply quality principles in the nonprofit education, health care and environment sectors of our community. An ambitious vision was created. Of course, writing a vision down is a bit easier than actually achieving one.

The second meeting was held in a large hall and each of us who had attended the initial team meeting stood beside an easel with the name of the focus area we wished to work on. New volunteers were asked to choose from one of six possible focus groups to join. The six focus groups were as follows:

1. Environment
2. Education
3. Health Care

4. Government
5. Manufacturing Industries
6. Service Industries

QNJ Environment Focus Group

I chose to lead the environment focus group for two reasons. First, as part of their coverage for the 1989 gubernatorial election, the New Jersey *Daily Record* ran a survey of New Jersey voters to identify the most important problems in the state.[7]

The following issues were perceived by the voters to be the most important:

1. Drugs	6. Taxes
2. Car Insurance	7. Gun Control
3. Education	8. Abortion
4. Clean Beaches	9. Mass Transit
5. Crime	

The environment (ocean beach water), was one of the most important problems to New Jersey residents, which made it worth working on.

The second reason I chose to focus on the ocean pollution problem was my summer vacation at the shore in 1988, when the beach was closed. My youngest daughter, Michelle, had told me that she didn't want to go back to the Jersey shore again. She wanted to go to Nana and Papa's lake house in New Hampshire, where they could swim in the water. Living in New Jersey, I wanted my kids to be able to swim at the Jersey shore again some day, while they were still young.

Slowly, several people gathered together around the environment focus group easel: one from Fort Monmouth (an army base), one from IBM, one from a small consulting company, and three quality managers from AT&T. We introduced ourselves and started our work.

First, we defined the problem: "New Jersey ocean water beach quality does not meet the expectations of residents." (We have a way of understating things in quality.)

Next, we discussed an approach: "Application of total quality management principles to reduce or eliminate the root causes of the problems." Elimination of root causes will prevent or reduce the frequency of problems.

To ensure a consistent approach, and because the focus group team members came from different companies and had different quality-training backgrounds, we agreed to attend a three-day training class at AT&T on AT&T's process quality management and improvement (PQMI) methodology. PQMI is a quality method for improving processes and addressing root causes of problems. I was going to be able to determine whether AT&T's quality approach could be applied to community problems.

Back at work, I called AT&T's environment and safety officer and requested his organization to "volunteer" a water quality expert for our team. We ended up with two volunteer experts on water quality, in addition to our small initial team.

Lessons Learned (1987–1988)

1. The causes of a serious problem are not always easily visible.
2. It is tempting to blame what is visible for all the problems.
3. It is tempting to blame someone you don't trust for being the cause of serious problems.
4. The actual causes of a serious problem are usually multiple in number, and difficult to uncover.
5. We can learn from the work of many people, from a variety of places.

CHAPTER TWO

The Pollution Process
(1989)

T he QNJ Environmental team's three-day training session was held at AT&T's Hopewell, New Jersey training center just outside Princeton. The training material was developed by AT&T's school of business and was based on a book published by AT&T's quality office. The training center was in a beautiful part of New Jersey, with deer from the surrounding woods running across the lawn, an appropriate setting for our QNJ environment focus group work to begin.

The course was intended to be experiential: the instructor explains a step in the method using a simple example, then the team attempts to apply that step to their own problem. The AT&T process quality management and improvement (PQMI) methodology, a quality method used by our company to improve key processes, has seven steps. The instructor explains each step and answers questions for about one hour. Then the team moves to a break-out room for two hours and attempts to apply that step to their own problem. Experiential learning is superior to the traditional continuous classroom instruction method of learning that most of us received in school because you immediately get a chance to see if you can apply what you've learned to a real-world problem that you are motivated to solve. It also provides an accelerated method, using dedicated time, to proceed through all the steps of the improvement method for a particular process. In three days, the team is trained and produces a fast prototype of potential process improvements.

A second team, from the AT&T quality office, joined us in the training and simultaneously worked on improving the AT&T quality office's administrative processes. After each break-out session each team reviewed their break-out work with the instructor and the other team. Each step took roughly half a day, allowing us to go through the first five steps during the three days of training.

The seven PQMI steps are as follows:

1. Establish process management responsibilities
2. Define the process and identify customers' requirements
3. Define and establish measures
4. Assess conformance to customer requirements
5. Investigate process to identify improvement opportunities
6. Rank improvement opportunities and set objectives
7. Improve process quality

THE SIGNIFICANCE OF PROCESS

Most people, particularly professionals or skilled trades people, learn how to do their work through education, vocational training, and on-the-job experience. Each day they do what they have learned from others, perhaps adding their own ideas on how to do a better job. Few people have documented their process for creating output. Few experts take the time to share their expertise with others, and when they do, not everyone has the time to learn.

When people work together on complex processes, such as a large building, managing the process becomes more important. For example, an architect (who designs the building and advises in the construction) and a construction foreman (who leads the team of construction workers) work together to design and implement a process that results in the successful construction of the building. In manufacturing, the engineer performs the function of the architect by designing the manufacturing process, and the shop supervisor leads the team in implementing the process for successful manufacture.

In the case of large and complex projects (buildings, the AT&T communications network, high volume manufacturing operations, or the sewage infrastructure), small defects anywhere in the process can lead to very large quality and cost problems. For example, if "800" number communications lines go down, customers can't reach a company to do business. A communications system defect can mean a loss of business.

While quality approaches have long been used in manufacturing, most service organizations (for example, airlines, banks, credit card companies) were slow to define and improve their processes. Those that have done so are now enjoying the benefits of higher quality and higher customer satisfaction at a lower cost.

The more people involved in creating the output, the more important it is to have a quality process, and a quality method for continuously improving that process. Because government is very large, a quality approach is appropriate to solve the large and complex problems that government faces.

Many government organizations are beginning to understand how quality principles can apply to them. One recent example is the reengineering of the U.S. Customs Agency, which was the first federal agency to develop a plan to implement the Reinventing Government program (a quality improvement program) that President Clinton and Vice President Gore announced in 1994. I was invited to Washington to advise U.S. Customs as they developed a process improvement plan. They were also the first federal agency to receive Congressional approval for their reengineering plan, a plan which promises to significantly improve their quality and reduce their costs. Going through customs could become easier for the honest people and harder for criminals.

Employees of companies can improve their processes if they are trained in a number of quality tools and methods. Therefore, the companies who do a better job of training their people develop a significant competitive advantage. A number of studies have shown that companies with higher quality levels produce higher customer satisfaction, market share, profits, and employee satisfaction. Some of these studies are referenced on the National Quality Program Web site: http://www.quality.nist.gov. Quality tools and methods are also beginning to be taught in our universities.

A process is "the way we do work to produce an output." A process is generally understood to include five common elements, often referred to as the Five Ms:

1. Methods (procedures or software)
2. Materials (raw materials and parts from suppliers)
3. Machines (computers or hardware)
4. Members (human resources)
5. Measures (operations and output measures)

In addition to these five Ms, which describe the common elements of a process, I have found another five Ms that are essential to the improvement of a process:

1. Market (driven from customer requirements)
2. Management (accountability and responsibility)
3. Motivation (positive and negative incentives)
4. Money (resources for maintenance and improvement)
5. Miracles (stretch goals based on benchmarks)

THE SHORE POLLUTION PREVENTION PROCESS

With this handful of M & Ms, we are ready to dive into our New Jersey ocean water quality problems. In our training session, we developed the following preliminary answers to the questions contained in the first few steps of the seven-step PQMI approach.

Step One: Establish Process Management Capabilities

Our initial discussion identified the following known responsibilities:

- The federal government (the EPA) is responsible for dealing with offshore dumping and oil tanker spills.
- The state government (the DEP) is responsible for the criteria used to measure the quality of New Jersey coastal water.
- The four counties along the New Jersey coast (Monmouth, Atlantic, Ocean, and Cape May) are responsible for measuring the ocean water quality in their respective counties and for reporting the results to the state.
- The 90 New Jersey shore municipalities along the Jersey shore are responsible for maintaining the condition of shore sewage and storm-drain infrastructure (pipes and sewage treatment plants) and for keeping our streets and beaches free from litter.

- Our upstream neighbor, New York City, is responsible for taking care of its own sewage infrastructure, old decaying piers, waste disposal, and harbor sludge removal.
- Businesses were responsible for meeting strict standards for waste-water quality, and beachgoers were responsible for picking up their own trash.
- We knew that we had a beach water quality disaster and a lot of finger pointing as to who was to blame.
- We did not know who had responsibility for the entire process required to prevent New Jersey beach water pollution (Figure 2-1).

Various levels of government each accepted ownership for a piece of the process, but there was no single owner responsible for the overall process and the results it produced. There was no owner responsible for the overall goals of the process, for analysis of the problems in the entire process, for prioritizing improvement efforts, or for allocating the limited resources (our tax money) to make the necessary improvements. No one took responsibility for protecting this natural marine resource from pollution, which was a major issue that the team would have to address.

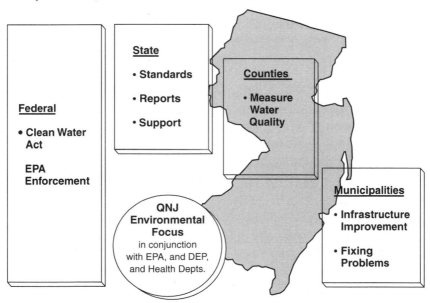

Figure 2-1. Parties Responsible for Pollution Prevention Process

Ocean Stewardship Responsibility and Accountability

In 1983, America extended its exclusive economic zone (EEZ) to 200 miles offshore by presidential proclamation. The U.S. EEZ is the largest of any nation in the world. More than 80 other nations now also claim a 200-mile EEZ. Within this EEZ, marine resources (both ocean and fish) are considered the property of the coastal nation. While this 200-mile EEZ was established to prevent over-fishing of areas by foreign fishing ships, it places a responsibility on each nation to play the role of trustee over marine resources for the benefit of all its citizens.

The ocean is a public resource that each nation has the responsibility to protect for use by the people of that nation. If each nation did this, the world's oceans would be protected.

The United States has no established measure of ocean water quality, no established measuring system, and no goals for ocean water quality. In contrast, all the nations of Europe have agreed to such measures and goals for the European ocean water quality as part of their European Economic Community agreement.

Because the federal government has claimed a 200-mile EEZ as its property, the public has a responsibility to ensure that the federal government manages this public natural resource in a responsible manner. The federal stewardship of this national marine resource is as important as ensuring that our national parks are not polluted.

This national responsibility is a critical one because more than 90 percent of all living marine resources in the world are now found within EEZs, and almost all fishing takes place in recognized waters of sovereign nations.[1]

Our U.S. government system was invented more than 200 years ago, long before we had a 200-mile EEZ, and before we had sewage treatment plants. Back then, New York City was quite a bit smaller and its waste and old piers were not a threat to the New Jersey shore. We created a system of government based on the principle of minimum interference from the next higher level.

Now we have a large and complex country and, while government has grown, it has not always changed enough to deal with the realities of today's large and complex issues, such as what to do with sewage from more than 200 million people in 50 states.

A Communications Network and Sewage Network Analogy

Like the AT&T communications network, our sewage network has grown large and complex. Both systems must be updated and expanded continually to handle increased capacities and to avoid the problems of overflow or blockage.

We want our thought and idea outputs to be delivered to exactly the person we want, immediately, anywhere in the world, with the same quality as if he or she were in the same room. We want our waste to be simply disposed of without negative impact on the environment. We also want both done at a low cost.

From 1934 to 1996, the cost of a five-minute long-distance call across the nation dropped by a factor of 10, while the quality has improved vastly. The cost of sewage disposal has gone up during the same period, while, unfortunately, the quality of the sewage system has become significantly worse.

Some of the same technology used to manage the AT&T communications network, which enables high quality at a low cost, could also be used for the management of the sewage network. For example, in 1988, New Jersey had not yet mapped the sewage infrastructure on an information system to facilitate rapid repairs when a break in a pipe occurred.

Process Owner Partnership

Our QNJ environment team was at step 1 in our process quality management and improvement methodology and we couldn't identify an end-to-end process owner.

Initially, our environment focus group did not fully appreciate the importance of having a process owner. Over time, our team became a business–government partnership and took on many of the process-owner responsibilities, as best we could, without direct authority or control over the resources required for improvement. We analyzed the root causes of the problems in the entire process, established process measures, established process goals, motivated others to implement necessary improvements, urged that necessary resources be allocated, and recognized achievements. Ideally, the process owner would do this analysis and have direct control over the allocation of the resources required to make the optimum improvements with the minimum cost. Our team knew how to do the analysis but did not control the resources. The government controlled the resources but did not know how to do the quality analysis. Together we had the makings of a process owner.

We found that concerned and involved citizens could form a business–government partnership and perform many of the process-owner roles. In fact, business–government partnerships may be necessary to ensure that key processes that affect environmental issues are managed in a quality manner. Our forefathers expected citizens to be actively involved when they created our form of government, a government by the people for the people.

Our environment focus group recognized that we had four levels of government that were concerned with, but not responsible for, the end-to-end process of preventing pollution of our New Jersey ocean water. Each level of government addressed its portion of the overall process, but without the benefit of an overall architect or construction foreman.

Each level of government wrote its own laws and regulations, sponsored its own improvement efforts and frequently blamed other levels of government, or business, for the poor results and high costs. Because no one accepted responsibility, or accountability, for the overall process there were no overall goals. Of course, we had provision to punish for noncompliance, via various laws and regulations, but enforcement is costly and does not directly produce the desired result: clean ocean water.

Accountability for Process and Results

A quality approach, using quality tools and methods to improve a process, attempts to be proactive by identifying root causes of prior problems and making process improvements to prevent or reduce the annual recurrence of similar problems. That is what a quality approach is all about. Punishing those associated with a problem assumes that an individual was the root cause of all the problems and that punishment will eliminate recurrence.

Quality experts have found that members (people) are generally responsible for only 20 percent of the problems in a process. The other Ms (methods, materials, machines, and measures) are responsible for the other 80 percent. In addition, punishment is usually not the best corrective action for the 20 percent of the time that people are responsible for the problem. Training is more likely to prevent recurrence of the problem when people are the cause. Hence, a government system that relies mainly on regulations, laws, and punishment as its approach does little to eliminate the recurrence of problems, and it usually produces a system that just barely meets minimal acceptable standards at a very high cost. Unfortunately, that is the approach our government has been using, which accounts for why our environment, along with a number of other areas, is at or below minimal standards. It also explains why we are almost five trillion dollars in debt and have prisons as one of our top growth industries.

To produce excellent results at low costs, corporate leaders must identify responsibility and accountability for end-to-end processes and results. Responsibility and accountability for end-to-end processes and results must also be implemented in government, with or without a team to help.

Unfortunately, along the Jersey shore in 1988, our team had not yet found a government process owner, or a government results owner. To be successful, I felt we would have to find one, or get someone in government to accept this role, but we decided to push on without a process owner at this point.

Step Two: Define the Process and Identify the Customer Requirements

The Pollution Process

Normally a business process is developed to produce an output, with minimum defects and costs, that satisfies customers. In this case the normal task was reversed. Our small environmental group needed to determine how to prevent the output of the pollution-producing processes.

Prevention is most efficient and effective when accomplished at the source, so we began to identify all the sources of pollution in the "pollution process." This simple idea, we found, was difficult to implement. It was exacerbated when a source of New Jersey's pollution problem was in another state upstream from us—New York. If the federal government had been implementing its responsibility and accountability as property owner and process owner, it would have ensured that these upstream problems were corrected, to prevent additional problems downstream.

To understand the process for preventing pollution, we first drew a block diagram for the process of producing pollution. The sources of pollution in our block diagram included household and business sewage, wrong connections or leaks in sewage lines, poor maintenance of sewage treatment plants, pollution that enters the storm drain pipes, and litter and garbage that reach the beach or water.

Our block diagram of the "pollution process" showed the blue ocean as the end point for the brown sewage lines (Figure 2-2). Sources of pollution showed up as brown sewage, black oil leaks, old gray broken piers from New York, and a variety of colors for the trash ranging from red cola cans to white cigarette butts. Our "pollution process" was more colorful than desired.

The process for polluting the Jersey shore was large and included many sources of pollution. The process for preventing pollution of the Jersey shore would have to eliminate, or significantly reduce, all of these sources of pollution. The truth is we were not trying to clean up the New Jersey shore, we were simply trying to stop polluting it. Prevention is what quality is all about. The good news is that the ocean is resilient. Like the human body, the ocean is alive and has the capability to heal itself, if treated nicely for awhile.

Customer Requirements

Before determining the customer requirements—what customers need and want—we first had to identify the customers. The team determined that the primary customer for a clean Jersey shore was the beachgoer, which included more than 50 percent of the New Jersey population.

We also recognized that there was more than one customer for a clean ocean:

1. The beachgoer
2. The people who fished for sport or business
3. The pleasure-boating enthusiast and boating industry
4. The shore tourism industry

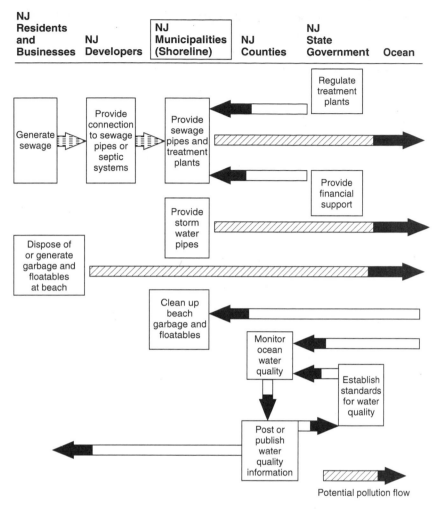

Figure 2-2. Defining the State of New Jersey's Process for Maintaining Quality Beaches

5. The 90 shore municipalities and 4 shore counties
6. The New Jersey state government
7. The marine life that lived along the Jersey shore

The beachgoer requirements appeared very simple. We had all recently been beachgoers. My experience was certainly memorable and it did not meet my expectations, or those of my kids. The beachgoer wants clean water and clean beaches that do not have to be closed due to pollution. The other customers seemed to want a similar result, for slightly different reasons. While we would have to confirm these customer requirements with a survey of beachgoers, our team, which was made up of beachgoers, was able to reach consensus on this simple customer requirement.

Step Three: Define and Establish Measures

In order to know if a process is producing the desired output, or results, measures must be established. In addition, measures at key steps in the process prior to the output—process measures—can be used to keep the process under control, thereby avoiding process output problems.

For example, a results measure could be the number of beaches closed due to pollution. Process measures might include the level of pollution in the storm drain pipes feeding into the ocean and bay waters, or the amount of floatables found in the New York harbor that drift down to New Jersey beaches.

To measure the process you need data. In our training class we had no data, but we could brainstorm process measures that could be linked to customer requirements. Because the beachgoers were interested in having a pollution-free bay and ocean for swimming, fishing, and boating, we tried to develop process output measures that indicated how well we were doing in meeting this customer need. A key process output measure we thought would be important was the number of days that swimming beaches were closed due to pollution each summer. We decided that the number of bays closed to the shellfish industry due to pollution would be another important measure. We added a third measure for safe boating, namely the number of boating accidents involving floatables in the water. Night boating could be risky with tons of decayed piers floating in the ocean.

Step Four: Assess Conformance to Requirements

Although we did not have exact data at this time on the beach closings during 1988, it was clear that the process for providing clean beaches did not even come close to meeting customer requirements. We subsequently found out that in 1988, New Jersey had the worst ocean water quality in the nation, based on state water-quality testing results reported by the Natural Resources Defense Council (NRDC) in its annual report on U.S. coastal water quality. The NRDC is a national nonprofit environmental group. The same NRDC report cited New Jersey as one of only several coastal states that tested and reported on the water quality of its entire coastline.

We also had a measure of customer satisfaction from the New Jersey *Daily Record* voter survey, done in 1989, that indicated ocean water quality had become one of the four most important problems in the state. Since that same survey cited taxes as the number one problem in the state, the solution to the pollution problems we faced would have to be accomplished without increasing government's costs. As quality professionals, we knew that a quality approach usually reduces costs while improving the quality of the output because it focuses on reducing the recurrence of problems, thereby reducing the cost of repair.

Though we knew that customer requirements were not being met, we were anxious to obtain actual data on beach closings for analysis of the process capability. We wanted to know if the process was capable of meeting beachgoer requirements and was simply out of control, or if the process was not capable of meeting customer requirements and needed to be reengineered. Bringing a large and complex process like this one under control would be a difficult management job, but reengineering it, a complete redesign and replacement of the existing system, would be very expensive.

We did a study at AT&T on the concept of quality in the U.S. culture, which indicated that we would rather reengineer things than simply improve things. Reengineering is sometimes necessary, but it is more expensive, takes longer and has a greater risk of failure. Over time, both process improvements and reengineering will be needed for every process, just like car repairs and purchasing of a new car are needed over time.

Quantitative analysis was needed to determine the strategic direction that the team would have to take to meet customer requirements. When we received data from the State of New Jersey Department of Environment Protection (DEP), we were able to determine that the process did not require major reengineering or replacement.

Our conclusion, after looking at the data, was that we had a very large process that was out of control, but one that could be brought under control to meet customer requirements with a reasonable amount of coordinated effort and expenditure. The next step was to identify process improvement opportunities.

Step Five: Investigate the Process to Identify Improvement Opportunities

The best way to identify improvement opportunities is to first identify the root causes, or sources, of the process problems, and then develop countermeasures. For example, if a problem is that my elbow hurts, the cause and potential solution is not obvious. If I have been learning to play tennis this week playing two hours a day, then the root cause of a "tennis elbow" makes the countermeasure or solution obvious. Stop playing tennis until my body can heal itself and the pain goes away. To "prevent" reoccurrence, improve my backhand, perhaps develop a two-handed backhand, and wear an arm brace. I have been through this problem, as have most tennis players. I am still wearing my arm brace.

The Root Causes of Pollution

There were quite a number of pollution sources of our ocean water that we discovered by using the many newspaper articles about our Jersey shore problems. Our list was expanded to include problems identified by our DEP team member, who joined our team after the initial training session. We identified the following 11 sources of pollution:

1. Pier debris—old piers in the New York–New Jersey harbor that had decayed.
2. Oil spills—from ships in the New York–New Jersey harbor.
3. Solid waste—dumping including medical waste.
4. Sewage sludge—dumping from sewage treatment plants in both states.
5. Sewage treatment plants—problems and failures.
6. Outfall pipes—Sewage treatment line outputs that were not fully filtered.
7. Storm drains—run-off including trash on streets of New Jersey and New York, animal droppings on streets, wrong connections

of toilets to storm drain pipes, leaks in sewage and storm drain pipes, fertilizers and pesticides from farms and yards.

8. Combined pipes—sewer and storm drain pipes coming into sewage treatment plants, with an overflow directly to waterways.
9. Trash from beachgoers.
10. Boater waste and toilets flushing into waterways.
11. Bird droppings near beaches and piers.

COUNTERMEASURES

The team then went through each of the sources of pollution and brainstormed countermeasures to eliminate each. These countermeasures became our process improvement opportunities.

Pier Debris

The old piers along the Hudson and East rivers were falling down, bit by bit, after each storm, resulting in tons of large pier material floating down the coast. These old piers were a risk to boaters and ended up as tons of litter on the New Jersey shore.

New Jersey had developed a program, using prisoners, to pick up and dispose of heavy pier remnants. However, until the clean-up could be done, the beaches were littered after each storm. More importantly, clean-up did not address the problem at its source. These decaying piers needed to be taken down before they fell down and floated to New Jersey. Prevention at the source would eliminate the beach litter and eliminate the risk of boating accidents.

As a result of the 1987–1988 Jersey shore pollution disaster, Congress authorized the Army Corps of Engineers to dismantle the decaying New York piers before they fell down, became floating hazards to boaters, and ended up as New Jersey shore trash. Unfortunately, Congress approved funding for only one year, 1989, leaving subsequent years as the responsibility of New York City.

Oil Spills

We knew that New Jersey, New York, and the oil companies that used the harbor had undertaken a major effort to stop the oil spills in the New York–New Jersey harbor. They had called in a professional consulting firm that used a quality approach to give them advice. The QNJ team decided to leave this pollution source for the oil companies to eliminate. A report on harbor oil spills, root causes, and recommended preventive measures was developed, and implemented by most oil companies. These improvement recommendations had the active support of both the New Jersey and New York governors, in response to the New York–New Jersey harbor oil spill crisis of 1988. Oil is very visible and therefore receives press coverage. Press coverage of a problem often leads to the political action required for improvement efforts. In addition, the oil companies created a Marine Spill Response Company which has the capability to clean up any spills, with resources supported by all the oil companies using the harbor.

Solid Waste

Throughout the summer of 1988, the New Jersey news media regularly displayed photographs of solid waste, medical waste including syringes, that washed up on the shore and beaches that had been closed due to pollution. The syringes were assumed to come from illegal dumping of medical waste from hospitals and medical centers. Controls at the source were needed, and they were put in place by New Jersey and New York health officers. Medical trash required extra controls at the source to ensure that it did not end up mixed with other solid waste and end up on our shore. Again, the very visible nature of this pollution facilitated both media coverage and government action.

Sewage Sludge

Sludge, the solid by-product of the sewage treatment process, had been dumped 12 miles offshore for 60 years, from 1924 to 1984. In 1984, the Environmental Protection Agency (EPA) concluded that this sludge dumping had damaged the coast. In 1984, the EPA picked a new dumping site for New Jersey and New York sludge, 106 miles out from the harbor and 80 miles from the beaches of Atlantic City. Our federal government was using its ocean property and process ownership rights to allow New York to dump off New Jersey's coastline.

The New York sludge dumping contract was awarded to the mother of a New York City political appointee, Susan Frank. Mrs. Frank reduced her dumping expenses by illegally dumping the New York sludge about five miles off the Jersey shore instead of the 80 miles required. Known as the Dragon Lady, Mrs. Frank was eventually convicted, and in 1995, was the oldest woman, at 80, in a New Jersey prison. In addition, all three of her children faced additional criminal charges for dumping sewage sludge into the New York–New Jersey harbor.

Another result of the Jersey shore pollution disaster of 1987–1988 was that Congress banned ocean dumping of sewage sludge in 1989. However, the ban did not take effect until 1992. New York was producing 225 tons a day of sewage sludge, which is about one ounce per person per day, and we had three more years of continued dumping to go. When the New York sewage sludge dumping was finally stopped off the coast of New Jersey in 1992, the difference in the water quality was observable. People could see their toes again.

New Jersey also had sewage sludge from its own sewage plants. The southern areas of New Jersey used landfills for their sludge. The northern areas dumped their sludge in the ocean. In 1988, the New Jersey sludge was being dumped 12 miles off the coast, but by 1989 and 1990, it was dumped 80 miles off Atlantic City. Starting in 1991, a year before the EPA ban on ocean sewage sludge dumping, all New Jersey sludge was sent to landfills.

Sewage Treatment Plants

For a variety of reasons, Asbury Park, a wonderful New Jersey shore town in the 1970s, went bankrupt in the 1980s. This was a case of a single shore town that needed state help, and still does, for the good of the entire New Jersey shore. The failure of the Asbury Park sewage treatment plant was the single largest cause of New Jersey beach pollution during the summer of 1988. The treatment plant was out of operation for three weeks, allowing raw sewage to flow directly into the ocean. The State of New Jersey worked with the counties and municipalities to ensure adequate sewage treatment plant maintenance. Legal action had to be used by the state to ensure that the Asbury Park treatment plant was operating properly. Ongoing support and help is still needed. The state should develop a plan to help towns in need. With this type of quality leadership, we could prevent a recurrence.

Outfall Pipes

When a sewage treatment plant's output pipe, the outfall pipe, does not terminate sufficiently far out into the ocean, under certain conditions the output can be washed back into the beach area. Because a sewage line output can be significantly above the fecal coliform levels allowed for swimming areas, this output could be another cause of pollution of swimming water.

One solution would be to improve the quality of the treatment plant output, but that could be very expensive. A more moderate cost solution would be simply to extend the treatment plant output pipe an additional 500 yards to avoid backwash of the treatment plant output to the bathing beach area. This is a workable solution for the four pipes in Monmouth County, where the original lines were engineered with a shorter length than elsewhere. Of course, implementation still comes down to each town or municipal sewage authority doing it. There are only 17 shore treatment plants in New Jersey and local citizens need to exercise their rights to protect our public beaches by ensuring local governments make such simple improvements.

Storm Drains

Storm drains and sewage pipes, running side by side, can crack at the same spot when there is a frost, resulting in leaks of sewage into the storm drain water system. In addition, in some parts of New Jersey, the sewage pipe system is constructed with 50-year-old clay pipes that had begun to decay. This decaying pipe system needed to be replaced, not fixed. Maintaining and replacing the sewage and storm drain system along the entire New Jersey shore involved an enormous network management job. The team later gathered data that showed it was being performed at various levels of performance by the 90 New Jersey shore towns. We needed to reduce the variation and improve performance all along the shore.

Streets needed to be cleaned regularly to avoid polluted run-off to storm drains. We found that many people thought the streets were cleaned just to have a clean street and they didn't realize that it was really to have a clean beach. As awareness of this root cause of shore pollution grew, in some towns kids painted pictures of fish on the curb near the storm drain. In other towns, kids painted the name of the bay, river, or wetland that the storm drain emptied into. This resulted in increased awareness of the real reason we need to keep our streets clean, and helped reduced trash in storm drains and the sewer systems.

We also needed help from the farmers. Farm manure piles needed to be moved away from locations beside rivers to prevent run-off during rain and storms. We knew the farmer did not want that manure pile beside the farm house, but the fish didn't want it beside their house either.

Combined Pipes

When it rains, the flow in the combined pipe is greater than the capacity of the treatment plant. The combination of storm water and untreated sewage overflows into waterways. In the New York–Newark metropolitan area, some sewer systems were originally under-engineered to save money, using one pipe for both storm water and sewage whereas two pipes are required to prevent pollution. There are about 750 of these combined sewer and storm drain systems in the New York–Newark metropolitan area. These systems overflow whenever we have a heavy rain, streaming pollution into rivers and bays. Unfortunately, pollution flows downstream. One of the team's silly ideas for a countermeasure was to move New Jersey upstream from New York, but the people of Connecticut didn't want to trade places with us.

A more reasonable solution was to install separate pipes for sewage and storm drain water, thereby eliminating raw sewage overflowing into bays and rivers from combined single-pipe systems. To do this required the addition of a second pipe for storm-drain water in all 750 such systems. We would have to reengineer these under-engineered systems.

Unfortunately, digging a hole and laying a pipe is viewed as too expensive of a project, particularly where the alternative is simply to keep dumping untreated raw sewage directly into our public waterways at "no cost." While this approach has a zero cost to the budget, it has an enormous cost to the public waterways. The Clean Water Act was passed to eliminate this type of pollution in 1972. However, the EPA continues to authorize permits for this source of pollution of our national coastal public resources.

The Clean Water Act of 1972 allowed the EPA to authorize states to issue permits for continued discharge of pollution into our waterways. All states have been authorized such permits. The EPA must draw the line on this continued dumping of raw sewage into waterways from combined pipes. Twenty-five years of permission to pollute, after the Clean Water Act was passed, is enough, but we do not yet have a plan for replacement of these combined pipes.

Trash From Beachgoers

To reduce litter on beaches and streets, a sufficient number of covered trash cans, for both beachgoers and pedestrians, needed to be made available and emptied daily. Easy to say, hard to do. This takes a continuous daily effort at the local level by every town. It can be done. It is being done by some. Where our public living areas are as important to us as our private living areas, citizens ensure that it is done.

Pump-out Stations

For boaters, a combined fill-up (gas) and pump-out (toilet) capability should be required at every gas station marina, to avoid the undesirable and time-consuming two stops generally necessary for boaters to fill up and pump out. Because many marinas have only fill-up facilities, some boaters simply pump out into our waterways rather than take the time for a second stop. Boaters must support the improvement of marine facilities required to prevent improper flushing.

Bird Droppings

Placing wire netting around piers near bathing beaches stops birds, particularly gulls, from sitting under the piers and prevents the bird droppings that pollute a bathing beach area. This is a simple and low-cost solution that does not harm the birds. In fact, it may improve the health of the birds by reducing their intake of junk food dropped on piers by tourists.

To summarize, the primary sources of the New Jersey shore pollution during the summer of 1988 were illegal New York sludge dumping too close to our shoreline; a number of oil spills in the New York–New Jersey harbor; the breakdown of the Asbury Park sewage treatment plant; floatable trash washed up on the Jersey shore from New York (old piers, medical waste, and trash); and sewage leakage into the untreated storm-drain water system. The breakdown of the Asbury Park treatment plant occurred within three weeks of when it was to be replaced by a new plant whose start-up was delayed.

DEPLOYMENT OF PREVENTIVE COUNTERMEASURES

At the end of three days, the team had ideas on how to solve these problems by identifying the root cause and preventing the problem at the source. Now, however, these ideas had to be implemented. Not just once, but every day of every summer season in all 90 New Jersey municipalities, and that little town to our north, New York City. How to achieve continued deployment of these preventive countermeasures every day, in every town, was going to be the real problem for us to solve.

Preventing Disasters

Out of disaster comes some good. The United States, and in particular our government, likes to react to disaster and we like to attack the big, visible problems. Oil spills in the New York–New Jersey harbor were being addressed, offshore dumping of sewage sludge was being addressed, although very slowly. Hospital and medical center trash was being controlled at the source, and decaying New York piers were being removed by the Army Corps of Engineers. The most visible of our pollution sources were being addressed, in one way or another, by a combination of federal, state, and business efforts, albeit a reactive response by government to the disaster of 1988.

There is a tendency in reacting to disaster to look for the one or two primary causes of the disaster, and address them. Then we like to think we have addressed the cause of the disaster and can go on with our lives.

However, I have found that disasters frequently result from the simultaneous occurrence of several small problems, while each alone would not have resulted in a disaster. For example, in the *Exxon Valdez* oil spill disaster, we had a ship's captain who was drinking, a government ship-to-shore communications system and shore-based spill facilities that were inadequate, and a ship that lacked a double hull. Before a disaster, each of these is viewed as a small problem, instead of as a contributor to a potential disaster that needs to be dealt with proactively. To minimize the number of disasters, we need a culture of continuous improvement on all small defects. By reducing the number of small defects, we reduce the probability of simultaneous occurrence of several small defects, which, when taken together, can lead to disaster.

While reacting to disaster is not bad, "only" reacting to disasters is bad. We need our government to proactively identify root causes of all problems and eliminate those root causes. Not just after a disaster, but all the time. A culture for continuous improvement of small problems, all the time, is the proactive approach needed to prevent disasters.

However, preventing disasters does not generate good copy for reporters and our political system is based on elections, whose outcomes are heavily influenced by media coverage. Most politicians are smart enough to know that reacting to disasters receives more publicity than preventing them.

Prioritizing Improvement and Assessing Customer Needs

The process owner normally ensures that the resources necessary to make the process improvements are made available. Typically, it does not take long to come up with more improvement ideas than anyone has the time, or money, to implement. Therefore, prioritizing improvement efforts, based on the overall costs and benefits, is essential to determine the best use of limited time and funds.

To prioritize our improvement ideas, and to produce the most improvement at the least cost, we needed data, which we did not have at our training session. We planned our next meeting and gave out assignments to collect the data that we thought would be useful to prioritize our improvement efforts.

Of course, without a process owner, we would have difficulty in aligning available resources with our recommended improvement priorities. We still had not found a process owner, someone who would reach across jurisdictions and ensure that the process was improved and that the results expected by the beachgoers were achieved.

We thought perhaps it should be the federal EPA, but they did not seem to want to manage the most important part of the process, which was the sewage and storm drain systems of the 90 New Jersey municipalities, and they did not seem to have a great concern for the quality of the New Jersey shore.

We thought perhaps it should be the New Jersey DEP, but they did not have any control over the EPA or New York and they did not have the resources to fund the required 90 shore municipality improvement efforts.

An additional data collection assignment was to survey New Jersey beachgoers on their needs and requirements. My secretary and I prepared the one-page survey and took responsibility for getting about 100 people to complete it. As a result of media coverage of the visible problems, New Jersey beachgoers mistakenly believed business, including waste disposal companies and oil companies, were the ones primarily responsible for the Jersey shore pollution problems. The beachgoers wanted zero beach closings due to pollution. The requirement for zero beach closings was the unanimous conclusion of the 100 people surveyed. We didn't need a larger survey to confirm the customer requirements.

DEPARTMENT OF ENVIRONMENTAL PROTECTION (DEP) JOINS THE TEAM

Our next team meeting included our initial eight members, the two engineers from AT&T's environment group, and my secretary. It did not take us long to realize that we had a lot of quality expertise in the room but still lacked sufficient data to analyze the root causes of the problems. We were brainstorming ideas, based on news reports of the problems, but lacked the facts required to act. Quality is based on acting on facts, not opinions. A lot of opinions exist; most are usually not fact based.

One of our AT&T quality managers proposed that we invite the state Department of Environmental Protection (DEP) to our next meeting. She wanted to ask them to share their data, and we would add our analytical capability. We decided to stop our meeting and call the DEP right then to invite a representative to come with data to our next meeting. After a few transfers, we had the state coastal water monitoring supervisor, Dave Rosenblatt, on the phone. Dave seemed a bit concerned about who we were, and what we were trying to do.

From his perspective, the typical "environment group" takes data and distorts it, or uses it to blame someone for problems. We tried to assure Dave that we were interested in analyzing the data, and using quality methods to help develop corrective actions. Our goal was to help the people in government solve the problem, not to blame anyone. Dave cautiously agreed to attend our next meeting and he informed us of an annual report he published containing data on beach pollution problems. He agreed to bring his annual reports to our next meeting.

This was the cautious beginning of the business–government partnership that would be essential to achieving our goal of excellent New Jersey ocean water quality. A partnership for continuous improvement was formed.

When our team started, we did not realize that we needed the government on our team, just as the government certainly did not realize that they needed citizens and business on their team. The government's sharing "their" data with an "outside" group usually resulted in the government being blamed later in the media for the problems. The idea of the public working *with* government to help protect a public resource seemed to be unique, and therefore suspect.

Assessing Blame

Democrats have said government has a responsibility to deal with our environmental and social problems in an effective way. They are right. Therefore, working with government for effective approaches to environmental and social problems is right. Blaming government workers for environmental and social problems is wrong.

Republicans have criticized government for being too big, costly and inefficient. They are right. Therefore, working toward smaller government and lower taxes is right. Blaming government workers is wrong.

Jim Florio and Bill Bradley, Democratic party candidates for New Jersey governor and U.S. senator, respectively, both made TV campaign ads at the Jersey shore that blamed business for the shore's pollution problems and promised to punish those responsible if elected. Both were elected in 1989. Blaming business was wrong.

We do not have to choose between efficiency and effectiveness in government. We can have both. We do not have to blame someone, government or business, for the problems. We simply need to learn how to solve the problems by finding the root causes and implementing the appropriate countermeasures. We can do this by applying a quality approach to key end-to-end processes, with a partnership for continuous improvement.

Teamwork Initiated

Quality New Jersey was underway. Our environment team was one of several QNJ focus groups and we had initiated the first QNJ business–government partnership for continuous improvement. We chose to focus on the environment, and the ocean water problem in particular, due to voters ranking this as an important priority. We had both quality and environment experts on our team and we had completed team training together. As a result of the experiential aspect of the training we had spent three days getting started on our analysis using the AT&T process quality management and improvement (PQMI) method.

We contacted the state DEP and gained their agreement to attend our next meeting and to share some data on the shore water quality problem. At our next meeting, we hoped to pull together sufficient data to complete step 6 of our PQMI method: *Rank improvement opportunities and set objectives.* This would be a critical step to help prioritize improvements given the limited resources for improvement.

Lessons Learned (1989)

1. Once you find all the root causes, the solutions are usually obvious countermeasures to the root causes.
2. The customer can provide a clear and compelling goal.
3. Experiential training can provide a good, accelerated start for a problem-solving team.
4. Clear responsibility and accountability for the results, and the process that produces the results, is required.
5. Some improvement can be obtained in the reaction to disaster, but reaction is not sufficient to prevent disasters.

CHAPTER THREE

God Made It Rain—
and That Caused the Pollution
(1990)

THE NEED FOR WORKING GOALS

In 1990, national ocean water quality goals for U.S. beaches had not been established, and as of this writing, they still have not been. While the European Community has established ocean water quality goals and standards for all of the countries in Europe, the United States has not done the same for the states. Without any goals or measures, it is hard to assess the impact of various state-by-state or national improvement efforts. A fundamental principle of the quality approach is that you need a goal and a unit of measure before you make improvements. Without a goal and a measure how will you know which improvement efforts are working? The goal and measure not only provide a check on improvement efforts, they provide direction and guidance to the broad base of people (federal, state, county, municipal, and watershed levels of government) that needs to make decisions and establish priorities on what improvements will be made.

ANALYSIS OF THE DATA

At our next QNJ environment focus group meeting most of our team met with the Department of Environmental Protection's (DEP) representative, Dave Rosenblatt. Dave was an environment specialist in his early thirties. He had been working on the New Jersey coastal water problem for several years and had quite a bit of knowledge about where the problems originated. While Dave did not have quality training he was interested in what we had to offer and was willing to learn. He was also very cautious about sharing government data on the ocean water quality problems because other groups had misused this data.

Dave's feeling, in 1990, was that many towns along the shore had government and business leaders that just hadn't thought enough about the importance of the ocean water quality to their community. They cared about tourism revenue and keeping local taxes down, but had not yet connected the need for excellent ocean water quality to these issues. They wanted to keep beaches open, but were not willing to spend the money on improvements necessary to prevent pollution.

At the local, state and national levels it seemed we had a long way to go, and New Jersey's 1988 beach pollution results, the worst in the nation, confirmed our concerns. However, Dave was interested in improving the situation and was willing to risk working with us and sharing his data on the possibility that we might provide him the help he needed.

Our first attempt to help Dave was to chart the data he had compiled on ocean water quality defects, evaluate the instances when a beach had to be closed due to pollution, to analyze trends, and to pin down the causes of defects. We also analyzed variations in the number of defects over time and over the geography of the New Jersey coast.

Standard quality analysis methods can turn raw data into useful, and truthful, information. We all know that data can be used to tell any story you want, if before you start there is not an agreed-upon approach as to how to analyze the data. Using a standardized methodology has the advantage of driving an improvement effort by fact, with analytical integrity, rather than by emotion with a predetermined bias. The facts also compel a team to come to consensus around the proper priorities for the improvements required, given limited resources. Without team consensus, quite often the wrong priorities are addressed, or nothing is implemented, while emotional arguments continue about which priority should be addressed. Unfortunately the political process, and our two-party system, falls into this trap quite often, with most of our legislators' energy spent arguing about priorities on a biased, emotional basis, rather than determining the appropriate priorities using an unbiased factual approach.

Our analysis of the trend of beach closures showed that, through 1988, the ocean water quality problem was becoming continuously worse. Without major improvement efforts the trendline projection indicated that ocean water quality would continue to get worse, not better. Our analysis of the variation in beach water quality along the shore, over time, indicated that the process for controlling pollution was out of control. This conclusion was drawn because the same location was not continuously producing the problem. The source of the pollution problem along the New Jersey shore was from a number of locations. Data analysis for different years showed different locations responsible for the poor statewide results each year. Although the widespread scope of the problem was initially discouraging, we recognized that the sporadic nature of the pollution meant that each location was *capable* of controlling pollution but occasionally went out of control. Out of control situations are usually reasonably inexpensive to fix, because the underlying process capability is there to do the job. It simply isn't being managed properly all the time.

Pareto analysis (a ranking of causes by the number of quality defects for which each cause or source is responsible) of the sources of pollution showed that the major causes were not the visible ones that the news coverage had focused on. The major New Jersey shore-water quality problem was not caused by the lack of national standards, it was not caused by New York's debris, it was not caused by business or oil spills, it was not caused by the state of New Jersey, it was not caused by the five New Jersey shore counties. The major source of the New Jersey shore pollution problem, which caused more than 95 percent of beach closings in 1988, was the sewage and storm-drain system along the entire New Jersey shoreline. Apparently the gradual decay and lack of regular repair and replacement of the sewage system was causing sewage to leak into the storm-drain system at a different place each year and flow untreated to our bay and ocean beaches. Part of the sewage system is the sewage treatment plant. The New Jersey shore municipalities have 17 sewage treatment plants along the shore, either run by an individual municipality, or run on a shared regional basis by a sewage authority on behalf of the multiple municipalities they serve. Individual municipalities with their own sewage treatment plant included Asbury Park, Ocean Township, Long Branch, Neptune Township, and Lower Township.

In addition, each of the 90 municipalities along the New Jersey shore had individual responsibility and accountability to maintain its own sewage and storm-drain system. We suddenly realized that we had identified 90 process owners for the major source of the ocean water quality problem in New Jersey (Figure 3-1). In 1988, roughly half the Jersey shore pollution problems were the result of a major sewage treatment plant failure in Asbury Park, and the other half were caused by pollution that got into the storm-water drain system.

When the Pareto analysis of the data showed that New Jersey's shore municipalities were primarily responsible for New Jersey's pollution problems, it was bad news and emotionally disappointing for me. I had just begun feeling good about having New York, our upstream neighbor, to blame for New Jersey's problems. It had sounded so logical, and we had lots of "evidence" from the media coverage indicating that New York contributed to our problems. However, the facts and data showed that the major contributor was not New York—it was our own Jersey shore municipalities. We were responsible for the vast majority of our own beach closings due to pollution.

As bad as the news was that New Jersey was responsible for its shore pollution problems, the good news was that the out-of-control New Jersey sewage and storm-drain system was in our state and we could therefore fix it by bringing it under control. It could have been worse. Had the system been under control and it was a low level of system performance that was causing the pollution, the solution would have required expensive reengineering or replacement of the system, perhaps costing multiple billions of dollars.

As it was, the majority of the pollution problem was caused by a system that was simply not being maintained adequately. For a reasonable amount of money, millions not billions, it could probably

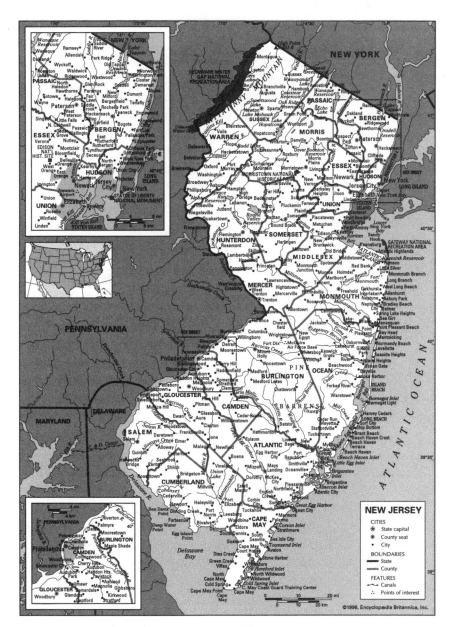

Figure 3-1. New Jersey Shore Municipalities Map

Source: Reprinted with permission from *Merriam-Webster's Geographical Dictionary,* 3rd edition, ©1997 by Encyclopaedia Brittanica, Inc.

be brought under control, using a good maintenance approach, thereby producing a substantial performance improvement at minimal cost. Dave was the only one not surprised by the results of our analysis because he had long felt that our cities were the source of the problem. However, he liked the "out-of-control" characterization of the problem that we developed from his data. This was a new way of looking at the problem for Dave, and it gave him hope that an affordable solution might be possible.

In 1989, the political commercials from our New Jersey candidates, as they stood on the Jersey shore, put the blame not on New York or the Jersey shore municipalities, but on business. If they were elected they would punish those businesses responsible for polluting our beautiful New Jersey shore. They would increase regulations and increase the enforcement of punishment for any business in New Jersey or New York that violated our environment. I knew from my quality experience that adding regulations and punishment would not improve results, particularly when the proposal was to punish those who were not causing the problem. However, it made a good campaign promise to blame someone who would then get punished if the right person were elected.

In 1987 and 1988, a handful of government officials and staff in the DEP thought that sewage and storm-drain system improvement was required and they had started work to write a Sewage Infrastructure Improvement Act. However, many believed that improvement would be prohibitively expensive. An example of this thinking that good water quality is not affordable was demonstrated in 1995 at the national level. The U.S. House of Representatives proposed and passed national legislation to circumvent much of the Clean Water Act of 1972 by attaching 17 provisions that either delayed enforcement or stripped funding from the enforcement of the Clean Water Act. The majority of representatives in the House believed that the cost of achieving clean water was prohibitive; therefore they voted not to have clean water. They were wrong. After hearing from many constituents, including me, the House reversed its vote three months later, thereby saving the 1972 Clean Water Act for now. However, with the EPA's "permission," much of the 1972 Clean Water Act has never been implemented.

The New Jersey sewage and storm-drain system includes 17 major sewage treatment plants with output pipes to the ocean to release the treated sewage water. What emerges from these plant pipes is not drinking-water quality. These pipes can produce a flow of pollution that drifts back to the ocean beaches under certain ocean conditions. Storm-drain pipes also flow out to the ocean with polluted storm-water run-off, and quite often they terminate very close to the beach. The New Jersey shore has 150 storm-drain pipes along the ocean beaches, each of which can be a conduit for pollution from run-off as well as from sewage system connections and leaks.

We have 7,000 storm-drain pipes that discharge into our New Jersey bays. The number of pipes that terminate in the New Jersey bays is much larger than the number that terminate in the ocean because it is cheaper to run a pipe to the bay from inland communities than to go the extra mile, or more, to the ocean. However, the bay has less ability than the ocean to wash pollution out to sea. Therefore, the combination of many more storm-drain pipes and less ability to wash the pollution out to sea results in the bays becoming more polluted than the ocean.

New Jersey is blessed with wonderful bays and inlets all along its coastline, which provide an additional 250 miles of bay and inlet shoreline. Each bay, of course, has two shorelines that New Jersey residents can use, as compared to only one ocean shoreline. Simple math shows that the potential value of the New Jersey bay shoreline is twice the value of the New Jersey ocean shoreline, however most people thought we needed to focus on saving the ocean shoreline because the bays were already so polluted that we might not be able to recover them. While we agreed to focus initially on improving the ocean beaches, which attracted most of the beachgoers, we also agreed to include the bays in our analysis and improvement plans.

MEASUREMENT AND TESTING

Dave explained to the team that the number of beach closings due to pollution was a function of how much rain we had each summer. The rain would flush the pollution in the storm drains out into the bay and ocean beach areas. Hence, the rain was one of the

triggers of the pollution on the beach by flushing the polluted storm-drain system. Since we expected it to continue to rain occasionally in New Jersey, we needed to enlist the infrastructure system owners in improving the systems.

The responsibility for maintaining the New Jersey shore sewage and storm-drain system belongs to 90 municipalities along the 127-mile New Jersey ocean coastline. Those who own the problem must take responsibility and accountability for fixing it, not others. In a quality approach, you are responsible for fixing the problems in your own back yard, in this case your own back bay. We were not looking for federal, state, or county takeover or funding; we were looking for those responsible for the problem. New Jersey's shore municipalities had to be held accountable.

Rain was not the cause of pollution but simply a triggering, a flushing, event. The ocean pollution was caused by storm-drain pollution that was improperly in the storm-drain system from a variety of sources. Professors at Rutgers, the state university, had begun to predict beach pollution by using rain forecasts and had determined a cause-and-effect relationship between rain and ocean pollution. When we had polluted beaches, news accounts blamed the recent rains. When we had no pollution credit was given to the recent good weather. This analysis confused the triggering event (rain) with the root cause (pollution in the storm-drain system).

Dave reported that, every Monday during the summer season, a number of points along the Jersey shore are checked for pollution measured by high levels of fecal coliform, an indicator for human sewage contamination. A high fecal coliform count is a likely indicator of potential bacteria and viruses. Of course, even when fecal coliform does not contain a lot of bacteria or viruses, I would prefer not to swim in it. If the fecal coliform level is above a level recommended by the DEP/EPA, the water is tested again on Tuesday. If it is again too high on Tuesday, the beach is closed. Each subsequent day, the ocean water is checked until the standard is met and the beach is reopened. The four shore county health departments perform these tests, share their results with Dave at the state DEP, and he then reports on the entire New Jersey shore's ocean-water quality based on these measures and a daily shoreline helicopter fly-over looking for floatable (debris or trash) pollution. This New Jersey

shore water quality testing program is one of only several in the country that is both comprehensive and results in beach closures when test results indicate a possible health safety issue.

I remembered my stay the previous summer on the New Jersey shore. Saturday and Sunday it rained, a triggering event. Monday, the weather cleared and we went in the water, which was not very clean. Tuesday, the second day of pollution, the beach was closed for the rest of the week.

When beach closure days are tabulated, Monday is not counted because it is the first day of testing and it takes two bad days of testing to close a beach. Using Monday as the test day was established to allow the municipality a chance to fix any causes of pollution during the week, prior to the high shore traffic on the weekend. Testing is important because it provides a trend in the level of defects. Analysis of the defects, found through testing, provides the data to identify the most significant causes of pollution.

To improve on the DEP measure of "beach closings," we created the measure of "beach-block-days of closures," a more specific measure that added duration and distance factors to our beach-closings measure. For example, one beach closure of three blocks for four days would be counted as twelve "beach-block-day closures" (3 beach blocks × 4 days).

To reach our goal of zero beach-block-days of closures, which was the goal that the beachgoers had told us they expected, we needed to eliminate the causes of pollution. The counties' testing once a week provided sufficient data to identify the sources of problems. We did not need more money spent on testing. We needed more money spent on improvements. If there was no pollution in the storm-drain system it would not matter when it rained. However, recognizing that pollution from a polluted storm-drain system would be flushed after a rainstorm, it seemed that testing after a rainstorm, in addition to every Monday, would provide a good check on the safety of the water quality. While the idea to add testing days was not a popular one in 1989, today some of the shore towns have begun to check their beach water after a storm and close the beach on their own if a problem is found.

In 1988, the state DEP reported several large-scale beach closings. However, these beach closings were of varying duration (from one day to three weeks) and of varying distance (one beach-block to several miles). In 1989, with less than half the season completed we were already over the number of beach closings reported in 1988. Nevertheless, Dave thought things had gotten better, not worse, from 1988 to 1989 based on his experience with the actual shore problems in both years.

The beach closings in 1989 appeared to be less severe in terms of duration and distance. In order to track our progress, we needed a measure that reflected a common understanding of the number of closures, the length of the beach closed and the duration of the closing, since all three factors directly affected the public's use of the beach.

Based on our work with Dave, the DEP reporting measure was changed to beach-block-days of closure beginning with 1989. A change to this measure indicated that 1989 showed improvement over 1988, while using the old DEP base measure from 1989 did not show improvement. We now had a more accurate measure of our shore water quality and it showed that Dave was right in his belief that 1989 had actually improved over 1988.

The shore municipalities whose beaches were closed in 1988 by the Monmouth County Health Officer later sued the Monmouth County Health Office for $20 million in damages due to lost business along the shore. Fortunately, this suit was dropped, but this legal action was indicative of the relationship between the municipalities, the counties, and the state during this period of severe pollution problems. The pollution led to a decline in New Jersey tourism revenue and the shore communities all felt it.

Some municipal officials recognized that pollution occurred after a rainstorm flushed the storm-drain pollution out into the ocean or bay water. These innovative public officials proposed that testing "not be done on Monday if it rained on Sunday." They had figured out that the results would probably be bad the day after it rained, so by not testing after it rained, their beaches could remain open. They were beginning to understand the problem, but not the solution.

Actually, testing probably should be done the day after it rains to ensure that people are not swimming in polluted water, and the beach should be closed immediately, to protect swimmers from possible disease if pollution is found. However, if we had pushed for this in 1989, we may have had a few more closed beaches but we would not have been a welcome member of the improvement team. Perhaps if my kids had gotten ill after swimming in the water on that Monday in 1988, I would have felt differently.

While the Monday tests didn't catch all the days that pollution exceeds the state limit, they did provide sufficient data to drive improvement efforts, providing we used the new beach-block-day measure of closings. And, over time, trending the results would indicate whether our improvement efforts were working.

WHAT TO DO WITH THE DATA

The testing data, and Dave's first-hand observations, confirmed that, in 1989, more than 95 percent of the beach closures were caused by pollution from polluted storm drains. We used our data on causes of pollution, combined with our understanding of the costs and benefits of various improvement opportunities, to develop a prioritization of the improvement opportunities that would produce the best results at the least cost, with responsibility identified for each opportunity (Figure 3-2).

While polluted storm-water run-off was a problem for everyone, the testing showed that some towns had perfect ocean water quality. It could be done. Perfect ocean water, no beach-block-days closed due to pollution, could be achieved in New Jersey and a few New Jersey towns knew how. These towns were keeping their pollution prevention processes in control, at an excellent level, all the time.

We decided that one role we could play would be to facilitate learning from the best, and sharing with the rest. We wanted all 90 shore towns to be as good as our best.

Benchmarking

This learning and sharing process is a common quality practice, called benchmarking, or sharing best current practices. However, many people don't believe that there is anything of value to learn from their peers, or don't believe that their peers' experience applies to their unique case. These beliefs inhibit the rate of learning in our society. It's a not-invented-here syndrome.

			Primary Responsibility		
A	**B**	**C**	**M**	**R**	**S**

	A	B	C	M	R	S
1. Improve water quality measures.						
• Measure beach water quality after storms.			x			x
• Measure storm water system water quality.				x		x
• Post or publicize results.			x	x		x
2. Schedule frequent cleaning at storm water catch basins.				x		
3. Remove storm water pipe discharge terminations from bathing areas.				x		
4. Ensure manure piles are not near waterways or storm water systems.		x		x	x	
5. Ensure adequately covered trash/garbage cans and frequent collection near beach and pier areas.		x		x		
6. Limit growth in areas of inadequate sewage system pipes (surcharge area), until pipes are replaced.		x	x	x		x
7. Require "best management practice" for storm water on all new building developments.		x		x		x
8. Find and fix sewage pipe leaks.				x		
9. Find and fix illegal and incorrect sewage connection to storm water systems.				x		
10. Replace any "combined" sewage/storm water pipe or collection systems.				x		
11. Require marinas to install and implement a "pump out while you fill up" capability.						x
12. Extend lines from treatment plants to an appropriate distance from bathing areas.				x		
13. Clear away decaying wood at pier locations, to prevent floatable debris.	x					
14. Stop disposal of syringes into toilet facilities.					x	
15. Develop and implement procedures to prevent oil spills.		x				x
16 Continue to utilize state prisoners to clean floatable debris. Expand to use county prisoners.	x					x

```
A = Army Corps of Engineers
B = Business
C = County
M = Municipality
R = Resident
S = State
```

Figure 3-2. Rank of Improvement Objectives and Responsible Parties

Motivating people to learn from others and to implement best current practices accelerates improvement. How to achieve this motivation is aided by an understanding of the quality archetype; that is, the cultural behavior of U.S. citizens toward quality. The U.S. cultural archetype was uncovered as a result of a study performed in the AT&T Network Systems Group, now part of Lucent Technologies, and documented in a book by Lew Hatala and

Marilyn Zuckerman called *Incredibly American*. This information is also included in one of our AT&T quality library books, *Quality Happens Through People*. Based on an understanding of the archetype, the system for motivation of the U.S. culture requires three phases that are aligned and coordinated.

First, there must be a clear crisis and perceived reason for change. The reason for change must come from forces outside the group that must change and it must be acknowledged by the leader of the group as a real and valid reason requiring major change.

The 1987–1988 New Jersey shore quality disaster was New Jersey's crisis: the worst ocean water quality in the nation. The team needed to remember this crisis and use it to drive the necessary change. Quite often, leadership would like to forget such a crisis and not mention it ever again. However, good leaders do not let people forget the crisis and use it to drive the improvements necessary to avoid a repeat of such a crisis. Our QNJ environment focus group not only created the beach-block-day measure, we recalculated the 1988 beach outage to be 855 beach-block-days. Each and every year, we published our results, including the 1988 results, so we would not forget the crisis.

Second, there must be new learning support for those trying to make the change. Our QNJ environment focus group shared with the municipalities both the root causes and best practices for eliminating the root causes of pollution. Each year we learned more from their successes and each year we shared more best practices. One example of support the state provided, as part of the Sewage Infrastructure Improvement Act from 1990 to 1993, was funding to shore municipalities and counties for mapping of their storm-drain and sewage infrastructure into a computerized information system. Mapping was the first step toward developing a comprehensive repair and maintenance program in response to our analysis that indicated a system out of control. This information system, along with county water testing results, could be used to quickly find and repair storm-drain or sewage-system leaks and misconnections. Prior to this, some people simply waited for the ocean to wash away the pollution and hoped it would not rain on Sunday, the day before the Monday tests! Another example of state support is the state DEP program using New Jersey prisoners to clean up heavy shore debris such as trees and lumber from decayed piers.

Third, there must be regular assessment of progress and recognition for those that make substantial improvements. Recognition based on quantifiable assessment of improvements made can motivate continuous improvement. Our QNJ environment focus group created a Shore Quality Award to annually recognize the New Jersey shore municipalities' and counties' achievements toward our goal of zero beach-block-day closures.

Most New Jersey coastal municipalities had much in common, such as old storm drains, reliance on tourism revenue and the negative perception of New Jersey's ocean water quality by potential tourists. They also had differences, including economic conditions that ranged from bankrupt to wealthy towns, population densities from three figures to six figures, beach frontage from only a few blocks on the ocean to several miles of ocean and bay beaches.

Our QNJ environment focus group, as part of our shore quality award process, helped by sharing our analysis of root causes with shore municipalities and by sharing shore municipality best practices for eliminating root causes of pollution.

AT&T—USING THE NATIONAL QUALITY CRITERIA

I was simultaneously implementing these same three phases at AT&T to motivate, support and recognize quality improvements by the business units and support divisions in AT&T. The AT&T units were like the New Jersey municipalities. Both needed motivation, agreed-to goals for excellence, education and learning about best practices, support for improvements, and recognition for their achievements. My experience in supporting quality improvements by the units of AT&T convinced me that a similar approach could work with the New Jersey shore municipalities.

On January 1, 1984, Ma Bell (the Bell System) was killed after having provided local and long-distance nationwide communications service for more than 100 years. To ensure that she was dead, the Bell System was cut into eight pieces. Seven pieces were each given a monopoly on the local telephone service in various regions of the country, the "Baby Bells." The eighth piece, the new AT&T,

courageously undertook entry into all the competitive businesses, which included microelectronics, telephones, PBXs (a telephone switch used in business), business data equipment, new computer offerings, communications network infrastructure equipment, and long-distance voice and data communications services. This new AT&T, in a fully competitive marketplace, was now my quality challenge.

Between 1984 and 1989, AT&T was not as fast and responsive as we needed to be in these very different market segments. We were still organized according to large, separate functional organizations, such as Western Electric for manufacturing and AT&T Bell Labs for research and development.

Beginning in 1989, Bob Allen, AT&T's new chairman, reorganized AT&T into about 25 product- or market-oriented business units, each of which had control of all the functional resources needed to compete in the various market segments. These business units were responsible and accountable for managing end-to-end cross-functional processes to achieve an excellent output at a minimum cost for our AT&T customers. Each business unit leader now had the ability to respond quickly to market changes in their different segments. In addition, AT&T retained about 20 support divisions that provided a resource, shared across business units when appropriate, to manage support processes such as law, real estate, public relations, finance and quality. This reorganization around business units was a key to AT&T's success during the next several years.

Unlike AT&T the QNJ environment focus group did not have a chairman to reorganize the large, complex functional levels of government into smaller, more manageable units focused on producing excellent results for an area of government, with end-to-end process-and-results responsibility and accountability. In the United States, the levels of government are like functional organizations. Each is focused on only a piece of an end-to-end process that produces results.

If government were organized like business units, I would have an environment business-unit head who owned the process *and* results for environmental issues for *all* levels of government. The environment business unit head would also have sub-business units for areas such as air, land, and water quality. Within the water quality sub-business unit we would have managers for ocean water quality, fresh water (rivers, lakes, and streams) quality, and drinking water quality. We would have someone responsible for producing results, not simply producing regulations for a piece of the process.

Instead, our team had to facilitate end-to-end process improvement by functional organizations at the federal, state, county, and local level, none of which had responsibility, or accountability, for New Jersey's ocean water quality results. Our QNJ environment focus group, in partnership with Dave Rosenblatt, the New Jersey DEP's coastal monitoring supervisor, became the environment shore-water-quality management team for New Jersey ocean water quality results. Dave was the line manager at the state level and the team helped him use a quality approach in providing the leadership required for all four levels of government.

To motivate AT&T's business units and support divisions to accelerate their rate of business improvement by learning and sharing best current practices, AT&T adopted use of the Malcolm Baldrige National Quality Award criteria in the form of an AT&T Chairman's Quality Award. This provided an annual management system assessment for the AT&T business units and divisions, and gave each organization feedback on areas for improvement, on an annual basis. The purpose was not to win an award, but to win in the marketplace.

The first year, we had 17 of our best AT&T units voluntarily apply for the award by having an assessment of their management system done using the Baldrige criteria.

In October of 1990, we held the first AT&T Quality Conference and invited leaders from leading quality companies to share their quality approach with us. We also planned for the first AT&T Chairman's Quality Award presentations, but unfortunately, our chairman had to deliver the difficult message that "we have no winners this first year because the unit heads set the standards high. While disappointing to all of us, we have established a good measure of our performance. I expect our units to improve and that some will meet our standards next year."

This was a tough message to give to the leadership team of AT&T. Some thought the reaction would be anger at those who managed the award assessment process (me). However, the assessments were accepted by those who participated, and concurred with by those that did not. We established credibility for the Chairman's Quality Award.

In 1990, AT&T also had several major network outages. The network is at the core of what AT&T is all about, since we made the equipment that went into it, and we operate and maintain it. The failures we had in the network had all in AT&T hanging our heads in 1990. We had let our customers down. The Chairman's Quality Award assessments confirmed that our performance level was not what it needed to be, unit by unit. More importantly, it gave us the feedback on unit-specific opportunities for improvement that we could act on to improve.

1990 was not a quality year at AT&T, and we were disappointed with our level of performance. But we had established a baseline of our performance against the rigorous national quality award criteria and the AT&T unit leaders went forward, motivated to learn from their feedback reports and improve from their baseline scores.

One of the major lessons from my AT&T and my New Jersey experiences is that deployment is harder, and more important, than the approach. Deployment produces results! Deployment must be motivated, not managed, and learning how to motivate deployment is the key to quality leadership. Those who know the approach play a very important quality support role, but quality leaders must know how to motivate the deployment that produces results.

The first year we ran our Chairman's Quality Award at AT&T, and concluded that we had no winners, was a very important step in motivating the quality deployment effort we needed. Motivation of improvement requires the leader to state clearly that current performance is not good enough and to set the standards expected. Our chairman had done that at our annual quality conference, and all of our unit leaders got the message.

AT&T'S PLAN-DO-CHECK-ACT (PDCA) CYCLE

The Baldrige criteria provide a good checklist for evaluating the quality of a business management system. In 1990, our AT&T business unit management systems did not meet the standard we had set for ourselves using this checklist.

To improve, we needed more than a checklist. We needed a quality approach to improvement and a deployment methodology for that approach. The quality approach that evolved from the work of our units was a simple plan-do-check-act (PDCA) model.

However, during the evolution of our PDCA quality approach, we had many units initially focus on only one or two aspects of the PDCA cycle without a balanced focus on all four elements. Some units liked to plan-check-plan-check. It took a while for them to realize that they needed to actually change some things to make improvements. Others liked to do-do-do-do without bothering to plan or check. They would simply jump to the next idea for improvement. Some began with hundreds of problem-solving teams that put them in an approach of act-act-act-act, only to find out that random action does not result in improvement. Those that took these partial quality approaches did not improve and began to feel that the quality approach did not work.

Quality management started for many companies on the factory floor and initially applied only to factory workers with a focus on the manufacturing quality of the product shipped. Total quality management (TQM) refers to engaging the total organization, all the functions and all the people, in implementing a quality management system to ensure that the total customer experience is of the highest satisfaction—from the design of the features to the service on billing questions.

For TQM to be successful, it should include both a total quality approach and total quality deployment. A total quality approach should include all the elements of the plan-do-check-act cycle.

At AT&T, the PDCA quality approach evolved from the best quality practices used by other companies and our best AT&T business units. Though there are many quality methodologies and tools available, we found the use of a total quality approach using the PDCA cycle provides the best support for continuous improvement.

The following few pages provide a description of AT&T's quality approach, the PDCA cycle, which also provided the basis for the shore quality improvement approach.[1]

Ready, Aim, Cycle

Aggressive strategic goals set the direction for AT&T; hitting the target requires improving the business, year-over-year, at a rate that consistently surpasses our competition. Hitting the target requires focus on what is important to ensure that people and resources were directly contributing to the execution of the strategy. Hitting the target will require the following steps:

- Alignment of all operations with strategic goals and objectives
- Design and management of key processes to achieve target objectives
- Rigorous analysis of performance capabilities and results
- Rapid and decisive action on improvement opportunities.

Hitting the target requires, in essence, continuous, disciplined implementation of the PDCA cycle (Figure 3-3).

The PDCA cycle is supported by AT&T's core set of quality methods:

Policy deployment is used to align operational goals and objectives with AT&T's strategic goals.

Business process management provides the framework for designing and managing excellent processes to deliver end-to-end services.

Management system assessment (AT&T Chairman's Quality Award) is a comprehensive check on the capability of the entire system to achieve world class levels of customer, people, and shareowner satisfaction.

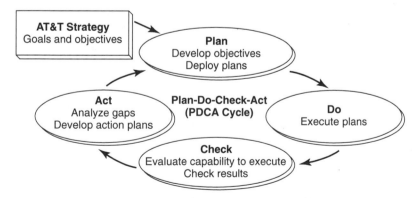

Figure 3-3. Hitting the Targets Through the PDCA Cycle

Problem-solving helps teams and individuals to analyze and eliminate gaps between performance and target.

Individually, these methods were used in many parts of the company to guide and accelerate change. But their power is realized only when the methods are integrated with the PDCA cycle and used along with rigorous analysis of appropriate measures to ensure that results are tracking favorably against goals. Taken together, these methods and measures comprise a quality approach for continuously improving at a faster rate than the competition (Figure 3-4).

A shared quality approach to planning, deployment, assessment, and improvement fosters teamwork and cooperation within and across all our operations, enabling process, management, and problem-solving teams drawn from across the business to meet on common ground.

Policy Deployment: Setting the Direction (Plan)

Policy deployment is a proven method for focusing on what is most important and linking operational capabilities with the key goals defining the strategic direction.

The fundamental characteristics that distinguish policy deployment from other planning approaches make it an especially

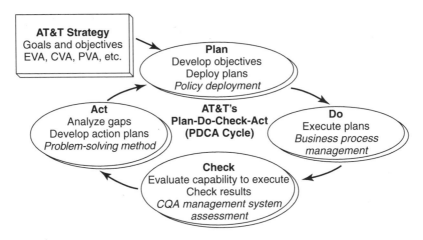

Figure 3-4. Integrating AT&T's Core Quality Methods With the PDCA Cycle

powerful method for planning and executing breakthrough improvements in the business:

- Annual objectives are established through a process of *catchball*, a term for give-and-take negotiation based on frank discussions of what is really required to get to where we want to be. In this catchball process, organizations derive objectives that are aligned with AT&T's strategic goals, directly linked with AT&T objectives and supported by implementation plans. Collectively, these objectives add up to the AT&T objectives.
- These vital few annual objectives, linked directly to the company's overall strategic direction, provide unambiguous criteria for making decisions and allocating resources.
- Clear measures and quantifiable performance targets associated with all goals and objectives clarify accountability, accelerate decision-making, and empower people to take effective action. *Measures* are the indicators used to track progress toward goals and objectives. *Targets* are set for measures based on baselines of current performance and benchmarks of competitors' and/or industry performance. The targets answer the question: How good must we be, by when, to achieve the objective?
- Objectives are *tracked*, and there is a disciplined schedule of review, in which resources may be reallocated and plans revised as appropriate, to ensure objectives are met.

Policy deployment's catchball process ensures wide communication and high visibility for the company's strategic long-term goals. This enables effective annual planning that promotes steady progress across the company year over year toward achieving strategic goals.

Basic Principles

Policy deployment supports the annual business planning cycle, as outlined below. It begins with the leadership team and extends to all AT&T operations.

1. *Develop draft company-level objectives*. The senior leadership team begins the process each year with an assessment of strategic direction, the unique focus that will give us our competitive advantage. They review the goals that define major dimensions of the strategic direction, looking ahead four to five years—the

vital few statements that describe how it will be when we achieve our mission and implement our strategy. These goals have associated performance targets, usually based on customer expectations, industry benchmarks, and performance levels required for industry leadership.

Next, they develop draft objectives, with relevant measures and performance targets that define—in clear, quantifiable terms—actionable, measurable interpretations of each goal. Objectives provide focus and clarify ownership and accountability in the short term (typically annual). AT&T refers to our highest level annual objectives as our "stakes in the ground."

2. *Deploy and link objectives across the company*. Senior leadership engages key AT&T leaders in catchball—the candid, iterative process of communication and negotiation—to align organizational goals and link annual objectives with overall AT&T goals and objectives.

   Catchball is not completed until agreement is also reached on implementation plans and the resources required to achieve goals and objectives. How will gaps be closed? What resources are committed? The agreed-to performance objectives are backed by implementation or business plans that align efforts within and across operations. There is closure on catchball only when these questions have been answered.

3. *Implement plans*. AT&T units analyze key capabilities and identify those key business processes (for example, billing or order delivery) where breakthroughs are required to meet targets. Appropriate resources are allocated, both people and financial capacity, to ensure success. Targets are set for process improvements or other initiatives that define what is required to meet annual performance objectives.

4. *Track results and support improvement*. Progress is monitored as necessary on key process improvements and other initiatives required to meet performance objectives. Based on regular reviews, leaders intervene when appropriate—reallocating resources and making whatever changes are needed to close performance gaps. Teams find and fix the root causes of gaps between performance and objectives. The process provides ongoing opportunity to correct problems and adjust actions.

5. *Cascade goals and objectives*. Just as AT&T leaders use policy deployment to link goals across organizations and ensure that the company's resources are aligned with its strategic goals, leaders

continue the process of goal-setting, communication, negotiation, and tracking within their organizations. They augment their AT&T-linked goals, objectives, and plans with goals, objectives, and plans that reflect the unique needs of their operations. They initiate catchball to deploy the full set of goals and objectives within their organizations—and beyond organizational boundaries, as appropriate. They require clear, agreed-to objectives and support plans, and develop shared criteria for directing improvements, making decisions, and allocating resources. The result is a "golden thread" that ties our processes and the day-to-day efforts of our people to the strategic direction of the company.

Business Process Management: Executing the Plan (Do)

Policy deployment is AT&T's core method for setting and deploying AT&T-level goals. *Business process management* is a powerful approach to building the capabilities required to achieve them. AT&T uses business process management to define our vital few key business processes—the critical activities required to execute our strategy and achieve our goals (for example, new service development and customer care) and to ensure that these key processes operate effectively and improve continuously. AT&T refers to its key processes as its "process framework."

The power of business process management lies in the application of the basic principles of *ownership, assessment,* and *continuous improvement* to the key processes critical to the execution of strategic goals:

- Leaders identify the end-to-end "chain" activities critical to achieving their goals and assign clear accountability for the performance and improvement of these vital few key business processes that cross organizational boundaries. The result is a well-structured view of the strategic operations of our business and clearly established ownership of their performance.
- The key business processes have quantifiable objectives that are aligned with overall policy deployment objectives. Associated measures track how well the processes meet these objectives. The measures help to answer basic questions: Who receives value from the process? What do they require? Are these requirements being met? Are these the right measures to encourage overall process effectiveness as well as functional effectiveness?

• The process owner, listening continuously to the voice of the end-user customer, ensures that the right work is done and engages the people working in the process to assess and improve how the work gets done. The process owner establishes explicit responsibility for improvement.

Business process management provides the architecture for integrating supporting functions and operations to meet customer requirements. It gives us a way to learn from experience and apply that learning to improve the vital operations of our business. As problems are found and eliminated, opportunities emerge to build greater capabilities—ways to do things faster, better, and more cost effectively than competitors. The underlying assumption is that the need and the opportunities to learn and change are unending, and that the basic principles of business process management—ownership, assessment, and continuous improvement—create the structure for learning and the incentive for change.

Implementing business process management—managing across organizations to produce results without direct control—calls for teamwork, commitment, and dedication.

Basic Principles

Business process management provides the framework for change—incremental improvement or transformation. To deliver optimum value and sustain competitive advantage requires ongoing application of its basic principles.

1. ***Identify key business processes***. Leaders determine the vital few end-to-end activities that will be managed as key business processes based on strategic goals defined in the policy deployment process. These key business processes include leadership processes as well as those—like product/service realization, order fulfillment, billing, and customer service, for example—through which they deliver value to customers. (AT&T defines these key processes in a document we call our "process framework.")

   Because key processes are so fundamental to the success of the business, the list of key processes tends to be constant. Yet, each year leaders reevaluate the list against current business priorities and conditions to maintain their focus on the processes most vital to achieving their goals.

2. *Assign process owners and establish process teams.* Next, leaders determine ownership for the key business processes, giving each owner clear responsibility and accountability for process performance. To meet this responsibility, the owner convenes a team, representing the basic functions of the process, to assess end-to-end performance and recommend changes. The team is *not* an added layer of management, but simply provides a structure for current functional leaders to work together, across organizational boundaries, to assess and improve end-to-end operations. The process owner empowers the team, acts on their findings, and resolves any cross-organizational issues that arise.

3. *Assess process performance and determine what needs to change.* The process team has ongoing responsibility to monitor performance and find ways to improve. They begin by identifying objectives: What is the process expected to do? They quantify its contribution to strategic goals and objectives. They investigate needs and expectations of those who receive value from the process. They look at benchmarks to discover the standard set by the best performers. The result is a clear statement of purpose, understood and agreed to by all involved.

   They define the process and measure its performance. They outline the basic steps of the process as it currently works. They list the major inputs and outputs of these steps and define how each contributes to meeting the ultimate objectives of the process. They measure performance—how well the overall process performs and how well the critical intermediate steps contribute to this performance. Because these process measures typically link directly to the attributes that define customer expectations, the measures are often referred to as direct measures of quality (DMOQ).

   Finally, the team uses performance data to assess the capability of the process. The team compares the current state (process performance) with the desired state (objectives and benchmarks). They identify gaps and opportunities and establish priorities for the changes required to close the gaps.

4. *Decide how to change.* Can the gaps be closed through incremental improvement of the existing process or does closing the gap require a whole new approach to the process? This decision will determine the degree of change, as well as scope and direction of the process team's efforts.

- *Incremental improvement*. Incremental improvements are specific changes to the existing process designed to correct problems—chronic areas of rework, cost, and delay—or to capitalize on opportunities—places where "good" performance can be built into a competitive advantage. The process team sometimes identifies and implements incremental improvements directly. When the root causes are not obvious, they often charter a problem-solving team to find root causes and develop solutions.

- *Reengineering*. When gaps are very large, the team's charter changes from fixing the existing process to envisioning a new process capable of achieving required performance levels. Closing major gaps demands creativity and out-of-the-box thinking to effect fundamental transformation. Such end-to-end reengineering involves both significant investment and risk, and often has a profound effect on the people involved in the process. The decision to reengineer must involve the process owner, who enlists the resources and expertise required to plan and manage the change.

- *Implement change*. The process team ensures effective implementation of their recommendations: They clearly document and communicate changes to all involved. They measure performance to assess whether objectives have been met, and they make any adjustments necessary to see that the anticipated benefits are realized. If recommendations extend beyond the scope of the process, the business process management team engages the leadership team in implementation planning and resource allocation. The process team also determines whether solutions developed in their problem-solving teams should be identified as best practices for replication within their process or possibly in other AT&T areas.

Management System Assessment: Checking Capability and Results (Check)

The Chairman's Quality Assessment (CQA) management system assessment process is AT&T's core method for checking end-to-end execution capability by comparing our business against standards of business excellence. Introduced in 1990 to encourage and support annual assessment in AT&T business units and divisions, the CQA provided feedback for improvement to between 20 and 40 organizations each year through 1995, enabling overall progress on average from "below sound" to the "excellent" range. To support the strategic direction of the new AT&T, the CQA refocuses on the effectiveness of the market-facing organizations in partnership with operational and support functions, to successfully deliver all AT&T services.

Used by the major market organizations of the company— business markets division, consumer markets division and international markets division along with associated operations—CQA provides an annual confirmation that plans and operations are well-aligned with our customers and our strategy, are well-deployed and well-supported by our processes, and that we are achieving positive and improving results when compared with competitors and benchmarks.

The CQA process is a check on the health of the business overall and provides feedback for high-level, systemic gap analysis and action planning. It is not meant to substitute for the more refined and specific checks that guide improvements at operations, process, and function levels. Major organizations are encouraged to seek a comprehensive check on their operations through other programs such as state awards modeled after the Malcolm Baldrige National Quality Award. Process reviews using a combination of ISO 9000 and Baldrige process evaluation criteria, as appropriate, ensure regular and consistent gap analysis and continuous improvement. In addition, participation in professional recognition and evaluation programs provide calibration and direction for some functional organizations.

Basic Principles

The CQA process is an objective, consistent, fact-based management system assessment: assessment teams represent a broad perspective on the business and the diagnostic criteria are modeled on the Baldrige Award criteria.

1. *Understand the management system*. Organizations review key aspects of the business critical to competitive performance—the leadership system, strategic plans and goals, customer- and market-related performance, product and service quality, financial performance, key business processes, information systems and

management, and human resource systems and management. This comprehensive review develops new insight, as well as a shared understanding of how the business is actually working, and illuminates immediate opportunities to close obvious gaps.

2. *Assess the management system.* Using comprehensive objective criteria for business excellence (for example, the Malcolm Baldrige National Quality Award criteria or other national quality award criteria if appropriate) teams compare the existing management system with a world class standard. This process yields a consensus view based on data and interviews with employees throughout the organization, about strengths and areas for improvement.

3. *Communicate feedback.* These findings are the basis of feedback: a numerical indicator as well as reports and presentations on strengths and areas for improvement. The verbal and written feedback provide a qualitative interpretation of the numerical indicator and a record of findings used for gap analysis and action planning. AT&T wide analysis of multiple assessments also serves at the AT&T level to identify broad gaps and common problems, as well as best practices to be shared across the business.

    The numerical indicator, comparing level of performance against a scale for world class performance, is reported as a range or band, identifying management systems as immature (below 400), sound (400 to 600), and good to excellent (over 600). The numerical indicator, along with key business measures provides a balanced assessment of the business.

4. *Act on feedback.* The feedback stimulates fresh insight about which areas offer the greatest opportunities to improve business capability. By incorporating this insight into annual business planning, organizations address not only what is necessary to achieve current objectives, but how to position themselves to accelerate overall progress and year-over-year improvement. This process has been most successful in businesses that have truly integrated the assessment process into their management system as the basis for identifying and closing gaps.

Problem Solving: Closing the Gap (Act)

Problem solving is a structured approach to identifying and eliminating the root causes of the problems that separate us from our goals. Because it guides both teams and individuals to find and fix underlying causes, it is the most widely used of our core methods. Problem solving is applied to large-scale investigations lasting several weeks, and it is applied in the course of a meeting—using its basic logic to probe a specific question at hand. Teams use problem solving in policy deployment to close the gaps between performance and objectives. Teams and individuals use problem solving as an integral element of business process management to investigate and pursue improvement opportunities.

With its broad applicability, simplicity, and powerful logic, problem solving enhances our ability to collaborate and build teamwork and partnering across the company:

- It provides a context, with a common language and shared approach, for teams to come together from any part of the business and immediately begin to work on solving the problem at hand and to easily communicate results for implementation in other situations.
- It encourages us to move beyond superficial fixes to fundamental solutions. Problem solving uses data and powerful tools for data analysis that help us to probe deeply into underlying conditions.
- It supports sharing of solutions and learning. When all AT&T people understand the basic principles of problem solving, any team's results can be easily communicated for implementation in other situations.

Basic Principles

Problem solving teams form and dissolve in response to specific business problems and opportunities. Depending on their findings, a team may complete all of the activities in the problem solving method from problem definition to tracking solution results, or—more likely—it may recommend solutions to process owners and others who have the responsibility and the resources for implementation and follow through.

1. *Define the problem.* The team begins with clear agreement on the problem. As straightforward as this seems, it is critical to a successful outcome. How a problem is defined to a large degree shapes the solution. For this reason, problem solving calls for careful formulation of a problem statement—based on an

understanding of the desired state, the current situation, and the gap that exists. It is helpful to have both a well-crafted problem statement—"Orders are late 20 percent of the time"—and a success statement that expresses the target outcome in measurable terms, if possible—"Improve on-time delivery from 80 percent to 95 percent."

2. *Analyze root causes.* In probing the causes of defined gaps, the goal is to continue asking "Why is this so?" until fundamental, actionable causes are uncovered. The team can begin by brainstorming, systematically capturing its collective insight in a list of causes, grouped by categories (for example, material, machines, methods, measures, and members). The team collects data and facts to test and validate its list of causes. Before developing countermeasures, the team needs to identify priorities within the list of root causes, analyzing relevant data and facts—not simply opinions and perceptions—to determine which causes have the greatest impact on the problem. Their objective is to discover the critical few causes whose resolution will generate the maximum value for the business. Agreement on priorities, based on fact and data, will help to direct efforts at the critical root causes and so produce observable progress in a short period of time.

3. *Identify countermeasures.* To identify countermeasures— ways to counteract the effects of each root cause—brainstorming is again a useful tool. The team uses its collective knowledge to generate ideas for solving the problem—encouraging creativity and building on ideas—without initially challenging the merit of any idea. Next, the team evaluates proposed countermeasures, looking for those that are both practical and powerful, grouping and editing related countermeasures to organize the output of the brainstorming session. To select the countermeasures it will pursue, the team asks the following questions:

- Which solutions are likely to have the greatest impact on the root causes?
- Is there a practical method for implementing the solution?
- How much will it cost? Is it feasible?
- Might there be unexpected, undesirable effects?

4. *Develop and implement plans.* From the list of potential countermeasures, the team develops an action plan that specifies the

who, what, when, and how of implementation. The plan identifies resources, responsibilities, timelines, and specific measures of success. The team—or those responsible for implementation—monitors deployment and tracks progress against the measures of success.

5. *Assess results.* Finally, the team, or its sponsor, evaluates how effectively the plan fixed the problem. It compares the results before and after implementation, using the same measures, to evaluate current performance against the target. It looks for evidence that the countermeasures have produced the desired effect, and, importantly, it looks carefully for unexpected, undesirable effects.

   If the target is met, the countermeasures are incorporated permanently into operations. If the desired effect has not been achieved, the team continues its analysis. If the same root causes persist, work is needed to develop new countermeasures. If the perceived root causes are affected, but the overall problem persists, work is needed to identify true root causes.

6. *Share lessons learned.* The team works with the process management team to determine if the solution is a candidate for replication in other similar areas. Sharing of solutions is vitally important in order to keep the finite resources of the business available for new initiatives, rather than dissipating them on rework.

SUSTAINING LEADERSHIP THROUGH REPEATED CYCLES

AT&T has long recognized the plan-do-check-act cycle as a powerful approach for improving business effectiveness, an approach that has enabled major accomplishments throughout the company. Today, as we face stronger competition and more complex challenges with a sense of urgency that often leads us to seek shortcuts, the discipline of PDCA has even greater value and importance. And as the cycle is our strength, the AT&T quality approach—PDCA plus supporting methods—is the vehicle for accelerating our progress: It directs our decision making and resources to the vital few actions needed to achieve our key performance objectives, it ties process improvement efforts directly to key objectives, and—through cascaded measures and targets—aligns and engages all of our people with the objectives.

APPLICATION OF AT&T'S QUALITY
APPROACH TO THE SHORE

There were some important differences between my day job at AT&T and my volunteer community project with QNJ, but the same quality approach and deployment strategy could be used.

At AT&T, I was coaching the quality support effort for more than 40 major organizations, each of which had a unit head and a quality director for support. I was helping the chairman to motivate more than 40 business unit and division leaders to deploy our quality approach. Also, I supported quality managers throughout the company in understanding our quality approach, in order to achieve an excellent management system that produced excellent business results.

My QNJ environment focus group needed to lead a quality support effort for the 90 shore municipalities and the five shore counties. The municipalities didn't have quality directors, but they did have city engineers who helped guide improvement efforts of each municipality.

In my QNJ project, I provided co-leadership with Dave Rosenblatt, in motivating 90 New Jersey shore municipality leaders, and five New Jersey shore county leaders to deploy our improvement approach. I also supported Dave in developing our approach to use quality methods to analyze root causes of problems and prioritize improvement efforts.

My approach to supporting quality leadership at both AT&T and the New Jersey shore included implementation of the PDCA cycle:

1. *Planning Support*: Establishing measures and goals for excellence based on benchmarks and customer needs and requirements.
2. *Doing Support*: Providing new learning and best practice knowledge transfer support to those responsible for process management.
3. *Checking Support*: Recognizing those that make substantial improvement toward the goals, or achieve the goals, to encourage continuous improvement by all.
4. *Acting Support*: Encouragement of support for improvements needed within key processes.

The First QNJ Quality Conference

QNJ had its first annual quality conference on November 8, 1990, also World Quality Day that year, although I am sure very few in the world knew about it. New Jersey's Governor Florio proclaimed November 8, 1990 as Quality Day in New Jersey and urged public and private sectors to join the Quality New Jersey (QNJ) team in making New Jersey a better place to work and live.

At our first QNJ conference, we created an opportunity to share the best quality practices of major New Jersey corporate quality leaders. Our speakers included the chairman of J&J, Ralph Larson; the executive vice president of AT&T, Vic Pelson; and the chief operating officer of New Jersey Bell, Dennis Strigl. We also invited the Department of Environmental Protection (DEP) commissioner as well as the commerce commissioner to share their thoughts, and to gain their support for our QNJ environment team improvement efforts and goals.

Monmouth County was the focus of the shore problem in 1988 due to the Asbury Park sewage treatment plant failure (Figure 3-5). In 1989, five sewage system leaks had caused a much smaller number of beach-block-day closings in Monmouth County. In 1990 Monmouth County had achieved a remarkable zero beach-block-closures.

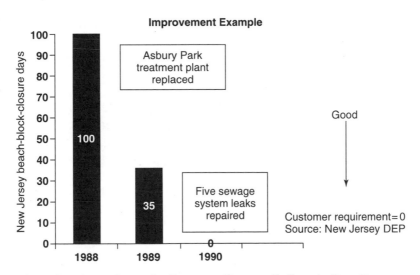

Figure 3-5. Investigate the Process: Monmouth County Results

This achievement was the result of improvements made by the shore municipalities and supported by the Monmouth County Health Office.

Our shore water quality was improving. New Jersey beach-block-day closings had fallen from 855 in 1988, to 276 in 1989, to 234 in 1990. Our QNJ environment team had set continuous improvement goals for the next six years: 150 in 1991, 100 in 1992, 75 in 1993, 50 in 1994, 25 in 1995 and 0 in 1996. The customer had told us that zero was the only acceptable number of beach closings. Our quality knowledge and experience in industry demonstrated that 30 percent improvement per year is achievable. Based on these two inputs, we set a stretch objective to achieve our goal of zero by 1996.

Lessons Learned (1990)

1. Measures of results (samples) are required to provide the data for control and improvement.
2. Data is required to prioritize improvements, given limited time and money.
3. Proper analysis of data can provide key information for improvement direction.
4. Trust is required, and must be honored, when data is shared. It must not be used to place blame, but to find root causes.
5. Teamwork requires measurable team goals.
6. Twenty percent of the root causes are responsible for 80 percent of the defects. It is critical to know which 20 percent, using data not emotion, and focus on it.
7. People will try to beat the measures, until you begin helping them beat the root causes of the problems.
8. All four elements of the PDCA cycle must be used to have an effective quality approach.

CHAPTER FOUR

Standards for Excellence
(1991)

AT&T STANDARDS FOR EXCELLENCE

Bob Allen, our chairman, was not yet an expert on the quality approach but he knew how to motivate quality deployment. He clearly stated the crisis we faced and with the help of unit leaders he set the standards of performance for excellence. He provided the new learning support needed and recognized improvements and achievements using the Chairman's Quality Award. Each year, beginning in 1990, Bob spent a day, or two, recognizing our units that had made significant improvements, or met the levels we had set for achievement.

We had no winners in 1990, but we had established a baseline of our current performance and had established standards for excellence. With disappointment in his voice, Bob delivered the difficult message—"We have no winners this year—we have not come up to the standards for excellence we have set for ourselves." He also communicated his expectations for improvement the next year.

NEW JERSEY SHORE STANDARDS FOR EXCELLENCE

Our QNJ environment focus group knew that the major causes of the shore problem were at the county and municipal levels, which meant we had to motivate deployment at those levels. So we created, as best we could, a process similar to that which we used at AT&T to measure, support, and motivate improvement.

To help us establish shore standards for excellence, we initiated a QNJ Shore Quality Award process in 1991, the year following the introduction of a quality award process at AT&T. We knew that the requirement for a 75-page Malcolm Baldrige-type application, covering all seven major category areas (leadership, strategic planning, information analysis, human resources, process management, operational results, and customer satisfaction) would be too much for the shore communities, so we created a simple three-page application. It covered only two of the seven Baldrige category areas: the key pollution prevention "processes" and shore water quality "results." This enabled the shore municipalities to easily fill out an award application, which our team could then judge. All applicants would receive feedback on how they did against the criteria that they could use for improvement. The best towns and counties could also receive statewide recognition at QNJ's annual quality conference, to help motivate continuous improvement. I sent a letter and the application to the four shore county health officials and the 90 shore municipality mayors, along with charts indicating the seriousness of the New Jersey shore water quality problems we all faced.

The application was designed to create an understanding, in each shore municipality and county, of the root causes and process improvements required to eliminate the shore water quality problems our team had found. For example, in the process section, we asked how frequently trash is picked up from the beach and what their process was for repairing storm-drain and sewage system leaks and for replacing sections of decaying sewage pipe and storm-drain pipe systems. In the results section, we asked for beach water-quality results trends, which we verified with the county data that the DEP had collected and reported.

We made it clear that we expected processes to be in place to prevent major causes of pollution. In addition, we communicated expectations for results trends that showed significant year over year improvement in their ocean water-quality measures—and/or achievement of a sustained level of zero beach-block-day closings due to pollution on their beaches. We were establishing standards of excellence for the New Jersey shore municipalities and counties.

Minimum Standards and Standards for Excellence

We only received eight or nine completed applications for the award our first year, but two met the team's criteria for an award, one municipality and one county. The winners were the small town of Avalon, in the southern part of the New Jersey shore, and Monmouth County, along the northern part of the New Jersey shore.

Avalon was one of the few New Jersey towns that knew how to maintain excellent shore-water quality—and they did it! Avalon had a perfect record for the past 10 years, with never a beach closure due to pollution, and it was not an accident. Avalon had processes in place to quickly repair or replace broken or decaying sewage pipes and storm-drain pipes, and they provided the budget to back up their program. They not only cleaned their beach daily but also cleaned their streets daily too, to ensure that pollution on a street did not have a chance to run off during a rainstorm into a storm drain that led to their beach water. Avalon was a small town—but they performed an excellent job. When it comes to keeping things clean, Avalon, New Jersey could be a benchmark for Disneyland. They became a benchmark for many of our other New Jersey shore municipalities.

We Don't Give Out Awards, We Give Out Fines

We had planned to present the first QNJ Shore Quality Awards at our QNJ annual quality conference in the fall of 1991. To motivate continued improvement by the winners, and increased participation by others, I wanted to have state leaders participate in the award presentations. Therefore, I called the New Jersey state DEP to request that the DEP commissioner participate in the presentation of these initial QNJ awards recognizing the excellent work of our first two winners. I was surprised by the reply I received from a DEP staff person, who told me that the DEP commissioner would not participate at our conference because "We don't give out awards—we give out fines. We don't want to confuse people by giving out recognition." Unfortunately, in 1991 QNJ was still a new organization that had not yet established credibility, and I therefore could not get through to speak to the DEP commissioner. Most likely, if I had spoken to the DEP commissioner directly, it would have resulted in a different decision by the DEP on the use of recognition.

I tried to explain that positive incentives can be used to encourage excellent performance at the same time negative incentives are used for poor performance. A leader can set goals for excellence and recognize those that achieve them. Setting goals for minimal performance and punishing those that fall below them is only part of the leadership job, and people won't be confused if you recognize excellent work and punish poor work. In fact, that is what most people have been educated to expect.

When government relies too much on punishing performance that falls below minimal standards it focuses people on achieving those minimal standards of performance, instead of standards for excellence. This is a key difference that can exist between how government and business manage. Business management uses both goals for excellence and minimal standards of performance. Both positive incentives and negative incentives.

I found this difference existed for two reasons. First, many people in New Jersey government still believed that higher quality meant higher costs and they could not afford higher quality. Some business people also still believe higher quality means higher costs and encourage government to set the lowest quality standards possible in order to minimize costs.

The achievement of higher quality *and* lower costs is possible through the application of quality methods to reduce the "cost of poor quality."

The costs of poor quality are the failure costs and repair costs that could have been eliminated by preventing the problem from occurring. For example, if New York removed decaying piers before they fell down and floated to the New Jersey shore, we would have had clean beaches *and* avoided the cost of cleaning up the shore. We also would have had revenue from beach use, avoided boating accidents from pier timber—and the cost of those accidents.

This first reason for lack of government leadership for excellent quality could be corrected with sufficient quality education of both business and government leaders. That education needed to include success examples, for those that will not believe it unless they see it. We hoped to create a success example with our shore team.

The second reason standards for excellence were not set was that people in the government felt they were only authorized to set minimal standards of performance, and they were not authorized to set standards of performance for excellence, as can be done in business. This thinking came from the idea that government's authorized role is to write and enforce the minimum rules of behavior (laws and regulations) for society. This role of establishing and enforcing minimum standards then leads to punishing (jail and fines) those who do not meet the minimal standards of behavior. I guess that is why we have a lot of lawyers in government leadership positions who think their sole job is writing and enforcing laws and regulations for minimal standards of behavior by business and society.

However, the role of leadership—in business or government—requires the setting of both minimal standards and standards for excellence. It requires both negative incentives for falling below the minimal standards as well as positive incentives for achieving standards of excellence. Writing and enforcing laws is only half the job of leadership. When only half the leadership job is done we do not have leadership, we have lawyering. Lawyers, or any other professional, can be leaders when they do both halves of the leadership job.

One national example of the government using recognition to motivate excellence is the Malcolm Baldrige National Quality Award.[1] The objectives of this award are to:

- Promote awareness of quality as an important element in competitiveness.
- Share information on successful strategies and benefits derived.
- Establish criteria for quality excellence.
- Recognize companies that excel in quality achievement.

This award, like our QNJ Shore Quality Award, is also the result of a government–business partnership to recognize, motivate and support learning that will lead to excellence in U.S. businesses and improve our ability to compete on a global basis. In the 1970s and early 1980s, the United States lost major industries, such as consumer electronics, and the jobs that go with them. We also lost a leadership position in other major industries, such as automobiles, due to our inability to compete with high quality and lost cost products, which some people refused to believe could be done, until it was too late.

U.S. businesses blamed their inability to compete in both cost and quality on a lot of things such as wage rates, tariffs, government support, and dumping. However, a few began to understand that there was a new way to produce higher quality and lower cost products—using a total quality approach. The Malcolm Baldrige National Quality Award was initiated in 1987 by a few who believed we should motivate the learning and deployment of a total quality approach in business to succeed in the United States as part of a global marketplace.

Enforcing Minimum Standards

One area where we spend far too much time and money is measuring to see if everyone is meeting the minimal standards of performance. Measurement can be done in a much more cost-effective manner by using a sampling approach to see if standards are being met. For example, government should not check everyone all the time to see if the minimal standards for minor things are being met, as we now do for minor house code inspections. We should have a random sample checked with a deterrent level of punishment if minimal standards are found not to be met.

For example, in Europe a "minimal standard of performance" is that riders are expected to have a ticket when they use the public transportation system. Monthly tickets can be purchased for reduced prices and people travel without each person having to go through a ticket validation check each time they ride the system. However, random sample checks of passengers for tickets are conducted and the punishment for riding without a ticket may be equal to the cost of a monthly ticket pass. In this case, the minimal standard of performance is having a ticket for the ride and it is checked on a random sample basis with a deterrent level of punishment if not met.

Toll booths are my favorite example of overchecking and overmeasuring. We stop everyone every several miles along a toll road going to the New Jersey shore, on every hot summer weekend, to collect 35 cents at each stop. This must be the most inefficient process approach that could have been developed to collect money to take care of our roads. I read an article that indicated that about 50 percent of the money collected must be used to pay the toll attendants, which means we are being stopped each time to actually collect about 17 cents for roads. A simple process improvement would be to raise the state gas tax by a few cents a gallon so that we could pay our road tax while we were already stopped. We would also eliminate the cost of the toll attendants and the cost of the time lost by all those caught in summer weekend traffic jams going to the New Jersey shore. The quality of the shore is impacted by the time spent getting there. This is another example of how to improve shore quality while reducing cost and increasing state revenue.

Our approach to regulation enforcement of minimal performance standards in all applications should be modified to reduce minor inspections. We should be clear about the minimal standards to be met, and use a random sample when necessary (as the IRS does) to check whether the standard is being met, and have deterrent levels of punishments if not met.

A quality principle is to reduce, but not eliminate, inspection in favor of prevention (for example, through education and upstream improvement efforts) as a means to improve the quality of the output while reducing the overall cost. For example, as minor inspections are reduced, efforts should also be increased to have people share best practices to meet cost and standards. The cost of checking is reduced and the savings are used to support improvement through planning and acting.

In fact, awards should be given to those who go beyond the minimal standards and meet the standards set for excellence. At the same time, deterrent levels of punishment should be given to those who fall below the minimal standards.

Citizen Involvement

At first, I thought government did not understand the excellence paradigm. However, I came to learn that most people in government believe their role, dictated by laws and regulations, is simply to enforce a minimal performance level and that they have not been authorized to set standards for excellence.

This is where the role and power of being a citizen in our country becomes important. I came to find out that those in government felt it was appropriate for a citizen group to set standards for excellence in a given area, and government could support, but not lead, that effort. Like the Malcolm Baldrige National Quality Award, my QNJ focus group was a joint government–business effort. It was business, or citizens, that had the right to set a standard for excellence, when government did not feel they had that authority. This is why citizens and businesses must be involved in our government. It was intended to be a government "by the people for the people." Without this involvement, there is minimal performance or, too often, below-minimal performance, at high costs. With citizen or business involvement, we can achieve excellent performance at low costs—using a quality approach.

In order for our current governmental process to produce excellence at low cost, instead of minimal performance at high cost, citizens have to be actively involved in how our government is run. In fact, citizens have to provide leadership for our government, because those in government do not feel authorized to set standards for excellence. As a citizen on my QNJ project, I was part of a team providing leadership for a shore improvement effort, whereas at AT&T, I was part of a team to support a business improvement effort.

QNJ SHORE QUALITY AWARD

Monmouth County was one of the first two QNJ Shore Quality Award winners in 1991. Monmouth County was also the county that had the worst shore water quality in 1988, in part due to a sewage treatment plant failure. Monmouth County is heavily populated and its storm-drain and sewage system problems, leaks, wrong connections, and decaying pipe had been perhaps the worst in the state in 1988.

The Monmouth County Health Officer who applied for the QNJ Shore Quality Award in 1991, Lester Jargowsky, was the same one who closed the beaches in 1988 and was personally sued by shore municipalities. He had since installed a countywide computerized information system that contained maps of all the storm-drain and sewage systems in the county, which was made possible by state-level funding because the DEP had convinced the state assembly and senate to authorize funding for mapping the shore storm-drain and sewage system.

A first step in a good repair and maintenance program is to know where your pipes are. At AT&T, we think it is important to know where our fiber optic cables are. Monmouth County was a leading county in using information system technology to drive a systematic repair and replacement program across the entire county. The county informed a municipality of indications of leaks or misconnections based on water-quality test results. They expected repair by the municipality of the source of the problem within 24 hours, thereby reducing continued pollution from unrepaired leaks. Monmouth County began acting like one of AT&T's national network control centers. Monmouth County began applying a quality approach and went from below minimal standards of performance to excellence in three years.

The following year, our QNJ environment team visited the Monmouth County Health Office, on one of our "field trip" meetings, to see their system in operation. We found they also had included the location of major toxic material sites in the county with an action plan, should a storm or accident cause a spill that would have to be cleaned up.

On this visit, we also became aware of a number of sewage pipes that crossed rivers and were pressurized to move the sewage uphill to the treatment plant. Most of the pressurized sewage river-crossing pipes were old and decaying, and a back-up pipe was not available. When they broke there could be an extended period of untreated pressurized sewage flowing into our rivers, and then into the bays and oceans, until new replacement pipes could be put in place. If we chose to put the new pipe in before the old one broke, a preventive program, we could avoid days of river pollution. If we waited until each old and decaying pressurized river pipe broke before repairing or replacing it, we could have a repeat of 1988, when major pollution was caused by one treatment plant breakdown.

The current funding did not support a preventive approach on the river crossings. We were still in a repair mentality. This situation was somewhat like a street corner where accidents happen and only after an accident kills someone do we spend the money to put in a traffic light. Perhaps, after one of the river crossing pipes broke and the pollution occurred, government would be convinced to fund a preventive approach to avoid repeated occurrences. On the other hand, if the goal is only to meet minimal standards, with the implicit understanding that even the minimum standard will frequently not be meet for reasons beyond our control—or beyond our budget—we may never take a preventive approach.

A sewage and storm-drain repair program, with short cycle times (less than 24 hours to repair) was generally good enough to produce excellent shore water quality. At locations like river crossings, where short repair-cycle times were not possible, a preventive strategy, a back-up replacement pipe, would be the appropriate strategy.

I had not yet had an opportunity to meet the 1991 DEP commissioner, our third DEP commissioner in three years. He was quite busy with the new job, combining the Department of Environmental Protection (DEP) with the Department of Energy to form a Department of Environmental Protection and Energy (DEPE) department. Being a lawyer with prior experience on the energy rate commissions, he was initially more involved with energy issues, ones he was more familiar with. This reorganization was a distraction, and dilution, of the focus we needed to improve New Jersey's environment problems. In a later year, it would be reorganized back to a separate DEP agency.

The 1991 QNJ Shore Awards were presented by me, alone, on behalf of QNJ. While I was concerned about the lack of involvement by state leaders in the presentation of the awards, the awards were still very much valued by our recipients. Lester Jargowsky, the Monmouth County Health Officer, was a strong six-foot man who had spent the early morning hours that day patrolling New Jersey bays. Early that morning, he caught people dumping toxic wastes in the bay. He was tired and irritated by the people he had been dealing with early that morning in doing his job. But when he accepted his award, a tear came to his eye as he explained that in his 25 years of work in government he had never been given an award before. This was a man who was sued for doing his job a few years before.

AT&T'S NETWORK PREVENTIVE STRATEGY

While I was facing problems on the New Jersey shore clean-up, at AT&T I also faced a number of problems during the 1990–1991 period. Some of the problems I found at AT&T in 1990 and 1991 were similar in nature to the problems I was encountering with our New Jersey shore water quality.

The years 1990 and 1991 were difficult for the telecommunications industry. We had our own outages—network outages. By the early 1990s, most of our business customers ran their businesses on-line; overnight data-center batch processing was becoming a thing of the past. However, the telecommunications industry network repair times weren't quite up to the demands of our information age business customers, which were similar to our New Jersey beachgoer requirements: "No outages, and if you have one, fix it in minutes not hours."

A telecommunications network will have outages; for example, fiber optic cables can be dug up by backhoe operators or cut by trains that fall off tracks. We already had a fast repair strategy for the telecommunications network and were now working on a very fast automatic restoration capability using spare back-up systems. We were simultaneously working to develop strategies to prevent backhoe dig-ups and trains landing on our fiber cable.

After one of my reviews with the AT&T chairman, Bob Allen and the AT&T senior executive for the communications services group, Vic Pelson, a change was made to the AT&T governance process to ensure that our network met our customers' quality and reliability demands. A full-time communications services quality officer, Frank Ianna, was appointed who would report directly to Vic Pelson and ensure that the quality and reliability capabilities of the AT&T communications network were continuously improved and given a priority in our capital budgeting process.

After Frank's appointment, I briefed him on several areas that required improvement based on a Baldrige criteria analysis I had completed. This analysis had identified areas that were root causes of outages. Frank undertook the responsibility for addressing the root causes of our network outages and for developing and implementing a corrective action plan. From 1991 to 1994, AT&T made a

20 times improvement in our network reliability. A factor of 20 improvement in three years!

Our customers benefited from the major investments we made in a fast restoration and preventive strategy, and AT&T achieved the most reliable service available to meet their demanding requirements. The ability to change the governance process, when it initially did not respond quickly enough to the customer requirements, was a key to our significant improvement.

THE MOST IMPROVED SHORE

The voters, beachgoers, and tourists in New Jersey also demanded no outages—no beach closures. The faulty sewage treatment plant was fixed. We had put in place a system that supported and motivated shore counties and municipalities to rapidly repair storm-drain and sewage system leaks or misconnections, which had been the major sources of ocean beach pollution. As a result, the New Jersey ocean beach-block-day closures due to pollution improved from more than 600 in 1988 to 10 in 1991—a great improvement in three years! While New Jersey also lost customers (tourists) in 1988 and 1989, by 1991 New Jersey tourism was on the increase and at a rate that made New Jersey tourism the fastest growing industry in New Jersey.

In 1991, the regional Environmental Protection Agency (EPA) office for the New York–New Jersey area, which also had a member on our QNJ environment team, established a short and long-term floatable action plan in conjunction with New York and New Jersey to deal with decayed piers and from 1991 through 1996 New Jersey had no beach closures due to floatables. This action plan collected floatables in the New York/New Jersey harbor before they floated down to the Jersey shore.

New Jersey's bays have almost 50 times the number of storm-drain pipes flowing into them than into the ocean. The bays' ability to absorb pollution is less than the ocean's due to the ocean tide's assistance in removing pollution from the beaches. In 1989, the bay beach-block-day closures were 232, but in 1991 we had 97, an improvement of about 60 percent. Good improvement, but still slower than needed.

Still, overall New Jersey shore (ocean and bay) beach-block-day closures dropped from 855 to 107 between 1988 and 1991, a factor of 8 improvement in three years. New Jersey had the nation's best improvement in ocean water quality. During this same period, California, Florida, and New York all had an increasing number of pollution-caused beach closures, with California and New York surpassing New Jersey's worst-in-the-nation 1988 performance. In 1991, California reported 1450 beach closures, according to the Natural Resources Defense Council (NRDC).

New Jersey is one of only six states with a comprehensive shore monitoring program. Twelve shore states have a limited program and nine have no monitoring program for their shore water quality based on the NRDC's 1994 report. While state monitoring programs are not all equal, the NRDC provides the only comprehensive annual report on U.S. shore water quality. In addition, the NRDC reports on the adequacy of each state's monitoring program. This is clearly a minimum standard function that the EPA should perform, but it does not.

In 1991, we began to see a payoff from our improved ocean water quality in New Jersey shore tourism. New Jersey shore tourism revenue was $12 billion in 1988 and did not grow in 1989 after the worst shore water pollution year ever in New Jersey. However, with the water quality improvement in 1989, 1990, and 1991, shore tourism was on the rise and hit $15 billion in 1991, growing at a rate better than any other industry in the state.

It became obvious that taking care of the New Jersey shore water quality was simply good business. The $3 billion in New Jersey shore tourism revenue growth, largely resulting from the improved water quality, provided an increase in state revenue of at least 6 percent of that (New Jersey has a 6 percent sales tax and an income tax that averages about 3 percent). In addition, it is estimated that the shore tourism revenue growth provided a growth of more than 50,000 tourism-related jobs in New Jersey, the largest job growth of all the industry sectors in the state. This job growth was equal to AT&T's total employment in New Jersey, and AT&T was the largest single private employer in New Jersey in 1991.

ALL BOATS RISE ON A RISING TIDE,
WHEN IT'S A CLEAN TIDE

Business tourism revenue was up, jobs and employment were up, state government and shore municipal government revenues were up, and the beachgoers' needs for clean beaches were being met.

With such a success from our QNJ environment focus group, the New Jersey commerce commissioner arranged a breakfast meeting with Governor Florio to brief him on the work of our group and the entire QNJ organization. The governor seemed pleased and interested in the ocean water quality improvement, and asked if our QNJ education focus group could help improve the efficiency and effectiveness of Rutgers, the State University of New Jersey.

QNJ: Education

Our QNJ education focus group had focused on the K through 12 level of education, with a quality effort that started with the superintendent of each school district. The group had completed pilot efforts in 12 school districts (supported by business contributions from the AT&T Foundation and a few other New Jersey corporations) and had developed plans to roll out a quality support effort to each of more than 500 school districts in New Jersey, based on lessons learned from the pilot programs.

To be responsive to the governor's request, the next year we formed a second education focus group, the higher education focus group. There are 42 higher education institutions in New Jersey and Rutgers, along with about 30 others, chose to participate in the QNJ higher education focus group.

In addition to the QNJ effort with Rutgers, because AT&T is the largest employer and Rutgers the largest higher education institution in New Jersey, AT&T and Rutgers formed a three-year (1994–1996) quality partnership that I was asked by AT&T executive vice president Vic Pelson to support. Part of this partnership included a full-time loaned executive, access to the AT&T quality office support materials, and access to AT&T executives and quality professionals on a volunteer basis.

1991 SHORE SUCCESS AND TELECOMMUNICATIONS
INDUSTRY STRUGGLE

As we finished 1991, the success of QNJ's environment focus group, which was focused on New Jersey's shore water quality, was creating the desired effect of encouraging additional applications of a quality approach in other key areas of state government. My volunteer job was having some success, but the job I was paid to do was facing a few problems.

As we finished 1991, the telecommunications industry was having quality problems in meeting customer demands for high network reliability. In response to this, AT&T changed its governance system and invested the capital required to make a substantial improvement in AT&T's network reliability. In the manufacturing part of our business, I had been encouraging and supporting one of our leading units to apply for the Malcolm Baldrige National Quality Award, and they had done so in early 1991. Now, later in the year, they were having some product shipment delays due to quality problems.

In 1991, we had 23 voluntary Chairman's Quality Award applicants, up from 17 the prior year, at AT&T. While none met our achievement award levels, four units earned an improvement award that we had established for improvement of 100 points or more over their prior year's baseline score (the Baldrige criteria use a 1000 point scale).

At the 1991 annual AT&T quality conference, we recognized improvement and encouraged improvement at a faster rate by all. Our 1991 QNJ quality conference had an aura of success, with significant shore improvements and the beginning of an improved government partnership on our effort to improve the quality of life and work in New Jersey.

Lessons Learned (1991)

1. Excellence is a stronger motivational goal than minimal standards.
2. The National Baldrige Quality Award criteria can be used to check on progress, provide feedback for improvement, and recognize excellence.
3. Positive recognition can motivate people to excellence.
4. Punishment does not ensure that minimal standards will be met.
5. Our government relies too heavily on laws, regulations and punishment.
6. Citizens must be involved to help set goals for excellence for our society.
7. Information technology used in business can be applied to help solve problems in government.
8. A fast repair strategy is required to achieve minimum performance standards, and a problem prevention strategy is required to achieve excellence.
9. Improved quality produces improved customer satisfaction, which results in increased revenues.
10. The governance process, as well as key operational processes, in business or government, has to be continuously improved to meet changing marketplace situations and new stakeholder requirements.

CHAPTER FIVE

"Once You Put Your Boots on, You've Got to Go All the Way"
(1992)

A SETBACK FOR NEW JERSEY

At the beginning of the summer of 1992, our QNJ environment focus group was confident of achieving continuous improvement based on our three-year success record in 1989, 1990, and 1991. Unfortunately, that was not to be the case in 1992.

In 1990 our governor, Jim Florio, a Democrat, raised New Jersey taxes after being elected in 1989 on campaign promises that he would not raise taxes. In 1991, people reacted by voting in a Republican majority in the State Assembly based on Republican promises to lower taxes, by reversing the 1990 Florio tax increase.

This time the elected officials kept their promise and in 1992 rolled back a state sales tax increase that Governor Florio had made in 1990. Governor Florio had raised the New Jersey state sales tax from 6 percent to 7 percent in 1990. In 1992 the Republican majority state assembly rolled the sales tax back from 7 percent to 6 percent, providing a 1 percent change that would result in an impact on our efforts to continuously improve the New Jersey ocean water quality in 1992.

When the state assembly reduced state income from taxes, they were not clear about exactly what should be cut on the expense side to keep a balanced budget. Unlike the federal government, the states are required by law to maintain a balanced budget.

A third of the way through the summer season of 1992, we had achieved an improvement in beach-block-day closures over the same period in 1991. Things were looking good for a fourth consecutive year of shore improvement. However, that is when the state DEP informed Dave Rosenblatt he could no longer continue on our QNJ Environment Focus Group, that he had to stop his work in collecting shore water-quality data from the shore counties and stop his work with the shore counties and municipalities on continuous improvement.

The Democratic administration interpreted the tax cut as requiring the elimination of the ocean water quality monitoring and improvement programs. Our team felt strongly that the ocean water quality monitoring and improvement was required to maintain and improve water quality. For example, the sewage system repair work was now driven by problems found from ocean water quality testing.

I called the DEP to talk to Dave and found out he had been reassigned to inspect sewage treatment plants. His supervisor would not release Dave's new phone number, to ensure that he did not attempt to continue working with the QNJ group on efforts to prevent ocean water quality pollution.

I called the assistant commissioner, Dave's supervisor's boss, and she confirmed the decision to discontinue support of the work with QNJ on a statewide prevention effort in favor of assigning Dave to inspection of sewage treatment plants. Though I tried to explain the benefit of a statewide prevention program over this inspection assignment, it was useless. The decision had been made, the budget cut would impact the New Jersey shore water-quality improvement effort that had been so successful over the past three-and-a-half years.

My tone of voice, I am sure, indicated my displeasure with the decision, with the treatment of Dave, and with the treatment of our QNJ team. I made it clear that continued state-level involvement and support was essential to motivating and supporting the continuous improvement by the shore counties and municipalities. A state-level shutdown of the improvement effort we had initiated, involving four levels of government, would lead to a reversal of the progress we had been making over the past three-and-a-half years.

Next, I called the DEP commissioner, who simply said that the state assembly was at fault for the budget cut, not the DEP. If I wanted to continue the shore water quality improvement, I needed to convince the assembly to reinstate the funding for the DEP coastal monitoring program that Dave had led. I talked and wrote to members of the assembly, who told me that the administration had the flexibility to determine which DEP programs were cut. The assembly told me it was an administration problem and the administration told me it was an assembly problem. It seemed to me that our shore improvement efforts were being tossed around like a political football.

During 1991, which was the best year for New Jersey shore water quality, AT&T was working hard on its own improvements. In light of the New Jersey setbacks at the DEP, I could not see why I should volunteer any more time trying to help the state when they did not seem to care. My time could be focused on helping AT&T improve its own business operations.

At this point, I called my team and told them that, without Dave and continued state support of our efforts, we could not make further progress. The business–government partnership on improvement of the New Jersey ocean water quality was over. There was no point in continuing our team. I was calling to let them know that the team's effort was being ended.

However, a few team members pushed back. It was their team and they weren't quite ready to quit. They requested at least one more meeting to make a team decision on what to do. Recognizing that I was dealing with a quality team, which meant empowered team members, I reluctantly agreed to their request for another meeting to discuss the situation and to make a team decision on how to proceed, if at all.

At the next meeting, without Dave of course, one of the members brought a quote from Ed Norton (played by Art Carney) who worked in the sewers on *The Honeymooners* television show.

Sharing some sewer philosophy with his buddy Ralph Cramden, a New York City bus driver, Ed said the guys in the sewers have a saying: "Once you put your boots on, you've got to go all the way."

The QNJ environment focus group appreciated the guys in the sewers. These were the guys we were helping to keep the ocean clean. Our team had put their boots on in this effort to improve the New Jersey ocean water quality, and they weren't ready to take them off until we met the goal the New Jersey beachgoers expected of zero beach-block-day closures.

One of the team members had worked on the election campaign for his state representative and felt we should take our success story to the legislative branch of state government—and try to get the assembly to clarify that the budget expense cuts should not come from DEP programs needed to improve the New Jersey shore water quality.

I agreed to call and write our state representatives requesting their support for the DEP programs needed to continue our efforts—including getting Dave out of a sewage treatment plant and back on our team. My calls and letters were to the co-chairs of the assembly's budget committee.

The initial legislative response to my calls and letters was "We didn't know we had cut programs that impact the shore water quality." In fact, the legislative branch still felt the executive branch had the authority to determine where the cuts took place.

The Republican assembly and the Democratic administration continued to blame each other for the cuts in the programs critical to the New Jersey shore ocean-water quality. I made follow-up calls and sent letters to the DEP commissioner, but no progress was made in resolving the problem.

When the counties and municipalities became aware that the state was discontinuing support for monitoring and improving the ocean water quality, some of the counties and municipalities also discontinued their monitoring and improvement efforts. In late July, as a result of the lack of a continued focus on the ocean water quality results, repair cycle times, and the root causes of pollution problems, we had our first increase in New Jersey beach closures, due to pollution, since 1988.

This crisis, in the middle of the summer season, received the most press coverage the shore had gotten since 1988, which was bad, of course. This stimulated the Republican assembly and Democratic administration to resolve their differences and agree that the DEP programs needed for our shore water quality improvement efforts could continue. The official authorization to continue funding the needed programs was signed into law on the beach with appropriate media coverage for both the Democratic and Republican government officials. Both parties took credit for saving our New Jersey shore water quality.

By the end of August, Dave was back on the job, but it was also the end of the summer season. Based on the number of beach-block-days of closures due to pollution, the summer of 1992 finished worse than 1991. For the first time since 1988, we had not reduced the number of beach-block-days of closures due to pollution. Our beach-block-days increased from 107 in 1991 to 111 in 1992. While this was not a large increase, we had had an improvement in the first half of the season and now faced a serious setback in our continuous improvement efforts. We needed to continue on a 30 percent improvement rate per year to reach our goal of zero closures by 1996. Over time, a 30 percent improvement rate will provide a company with a significant competitive advantage (Figure 5-1). Although we had no competitors in cleaning up the Jersey shore water, we wanted to hold ourselves to the highest achievable improvement rate. Unfortunately, instead of a 30 per-

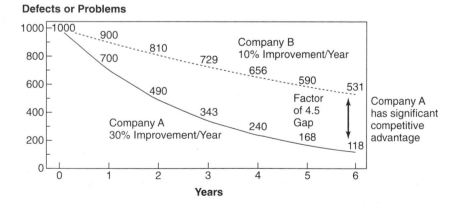

Figure 5-1. The Power of 30 Percent Compounded Improvement Rate

cent improvement in 1992, we had a 4 percent increase in pollution-caused closures.

The ocean beach closings increased from 10 to 27, while the
bay beach closings declined slightly from 97 to 84. With only 100
days in the summer season, these results were not good. We still
had an average of more than one beach-block closed due to pollution for each day of the summer season.

When the summer of 1992 turned into a setback for our
improvement efforts, our quality story and results did not look
good anymore. I was not looking forward to telling our story at
the upcoming annual QNJ quality conference in October. We had
nothing to be proud of this year, and the results raised some questions about our effectiveness and about the state–business partnership we had tried to establish as a role model for other QNJ
focus groups.

In October, I shared our story, explained the setback, and recognized two more shore towns that had excellent results in spite of
the setback on the state level. My QNJ associates understood. They
didn't question the efforts of my team, but simply encouraged us to
continue. Setbacks are simply part of life. The important thing is
not to let them stop our efforts for continuous improvement.
Without the QNJ environment team and the support of the entire
QNJ organization, I would have quit.

While 1992 was disappointing, at least we ended the year by
getting Dave out of a sewage treatment plant and back on the job
leading the state-level prevention effort. The QNJ environment
focus group, as a result of having faced a setback, was stronger
and even more committed to continue. We now had both the
state administration and state assembly aware of the needs of our
improvement programs to continue toward our goal of zero
beach-block-days of closure due to pollution. As a result of this
increased awareness, we felt the budget would not be cut again
next year.

We also had a new motto for our team: "Once you put your
boots on, you've got to go all the way." Dave still has this quote on
the wall in his office. The team still recalls it each time we face a
new setback.

1992 QNJ SHORE AWARD WINNERS

The two New Jersey shore towns that won our QNJ Shore Quality Award in 1992 were Stone Harbor in Cape May County and Stafford Township in Ocean County.

Stone Harbor had extended its storm-water run-off lines a greater distance from the beach so that possible street-water run-off that contained pollution would not affect beach water quality. They had also committed over $200,000 per year for their sewer improvements, a significant and sustained amount which was necessary to ensure the proper sewer infrastructure upkeep. Like Avalon, Stone Harbor conducted daily beach and street cleaning—using their neighbor Avalon as a benchmark of performance. At our 1995 annual QNJ quality conference we had a guest speaker from Disneyland to share their total quality approach with us. Their focus on daily street cleaning reminded me of the Avalon approach to keeping their city clean.

Stone Harbor, like Avalon, had a record of perfect beach water quality over the prior 10 years—no beach closings due to pollution. Stone Harbor is located as a neighboring town to Avalon and not only learned from Avalon but also minimized their costs by sharing the clean-up support resources, such as street sweepers, with Avalon.

Stafford Township was the second 1992 QNJ Shore Quality Award winner. Stafford had almost 50 miles of new sewer system installed or under construction. All of Stafford Township's beaches were on the bay side, which we knew was more difficult to keep clean due to the number of storm-drain pipes terminating in the bay. Stafford had developed a unique approach to their problem. They decided to reduce the number of storm-drain pipes that terminated in the bay. They created a natural ecosystem as a filter. They then terminated the storm-drain pipes in the ecosystem, which provided a natural filter for the storm-drain water, which might contain pollution from street-water runoff. Then the filtered water was released into the bay. This was a terrific idea that could also be deployed by some other shore towns to prevent pollution of our bay water.

I had learned my lesson from the DEP's negative view of recognition in 1991. In 1992 I simply invited the DEP commissioner, Bob Shinn, to make a few remarks at our annual QNJ quality conference and said nothing about recognition. I scheduled Commissioner Shinn on the agenda just before the presentation of the QNJ Shore Quality Awards. When he finished his remarks, I asked him to please stay on the stage with me while I called up the QNJ shore award recipients. He obliged, shook hands with the winners and posed for pictures for the press. The state leadership was involved and supportive and I was sure the DEP staff would now also be supportive.

KNOWLEDGE TRANSFER

Interestingly, while our team had been focused on preventing the sources of pollution, both of the 1992 award winners focused on preventing pollution at the source by moving the pipes that contained runoff pollution farther away from swimming beaches. I am always surprised at how much can be learned from others, particularly those in the field doing the day-to-day job, even when the problem has been studied by smart people for years. It is possible to get trapped into thinking that you have thought of everything—when a few people can never think of everything, particularly if they are not the people doing the job, but only people, like staff or consultants, analyzing the problem from afar.

The award winners were an example of why a few smart people cannot make all the rules or develop the best practices. Staff people need to establish a system for supporting the transfer of knowledge, rather than trying to be the source of knowledge. We do not need the smartest people in key staff or government positions. We simply need people willing to support a system that transfers knowledge from people who are on the line and continuously learning from their experience. That transfer of knowledge can be accomplished by several means which the staff person should support. For example, a transfer of knowledge can be accomplished by publications, such as newsletters, electronic shared folders, shared databases, books, courses, and consulting one-on-one with line people. As the quality office vice president at AT&T, I was able, due to our resources, to support all of these means of transferring quality knowledge throughout the company.

From 1989 to 1995, the AT&T quality office published more than 25 AT&T Quality Library books, each on a particular quality method and each containing the wisdom of AT&T line people, as well as those from other leading companies. From each of these books, I also had a course developed, which in most cases had gone through the process of being certified for college credits. Almost 10 percent of AT&T employees are taking a continuing education program to obtain a degree. Accrediting courses allows them to get credit toward their degree when taking the AT&T quality courses. I also was fortunate to have an excellent shared resource of internal quality consultants in what was then the AT&T Bell Labs. All of our AT&T units could contract with Bell Labs to support their own unit quality professionals when necessary.

When it comes to the transfer of knowledge, it takes all of the above resources to accomplish the task, and if we want to be a learning organization it is a job that is never done. We keep developing new wisdom from the experiences we have, along with the experience of others, and keep working on the transfer of those new insights. Over a number of years, the organization that has an effective support function for the continuous transfer of knowledge can develop a significant advantage over competitive organizations that do not, even though we all start with people selected from the same pool.

In my QNJ position, resources were more limited, but we shared our AT&T quality books and courses with the state of New Jersey, along with my QNJ environment team's quality consulting support. In addition, we communicated to all the shore mayors and county health officials our analysis of the root causes, a Pareto analysis to prioritize the problem areas, and the improvements being made by others. This knowledge transfer helped individual shore towns learn from each other the root causes and potential countermeasures to sources of pollution. In addition, it helped the four levels of government involved to work as a team, focus on a process, analyze data, and support the improvements required to achieve a goal.

QUALITY PROGRESS AT AT&T

Although we faced a setback on our New Jersey shore quality project in 1992, in early November I received a pleasant sign of progress on our quality work at AT&T. Two AT&T units won Malcolm Baldrige National Quality Awards and AT&T was the first company to have two units win the Baldrige Award. AT&T Universal Card Services won in the services category. AT&T Network Systems– Transmission Systems (now part of Lucent Technologies) won in the manufacturing category.

I had worked closely with both units over the prior three years. Both had participated in all three years of our AT&T Chairman's Quality Award assessment process in 1990, 1991, and 1992, where they demonstrated continued improvement of their management systems. Although these two units represented less than 15 percent of AT&T's total revenue, it was a very positive sign for the power of the AT&T quality office support processes that we had created. Both unit heads acknowledged the key role the AT&T quality office played in providing assessment, motivation and learning support for their quality improvement efforts.

Representatives of each of the two winning units were invited to Washington, D.C. in December 1992, to accept their awards from President George Bush. The winning units had a very limited number of tickets for guests at the award ceremony, but they invited me along. Their invitation was a particular honor for me; typically corporate staff people do not receive recognition from the units. I was also asked to provide some introductory remarks for a live video satellite broadcast of the award ceremony that we broadcast back to all 12,000 people in the winning units' locations.

AT&T'S NETWORK QUALITY

In November 1992, Jane Redfern, the quality director for our largest business unit, AT&T's consumer long-distance business, Consumer Communications Services, approached me for advice and counsel on their applying for the Baldrige Award (MBA) in 1993. AT&T had about 25 business units and only one applicant per company per year was allowed to apply for the award in each category—service and manufacturing.

The AT&T process for determining if we would have a Baldrige applicant in either, or both categories, each year was based on AT&T's Chairman's Quality Award process, which used the Baldrige criteria. Based on the 1992 CQA results, I encouraged Jane's unit to apply for the Baldrige Award in 1993, but to apply in conjunction with our AT&T network services division, the part of AT&T that designs, installs, and maintains the long-distance network for both our consumer and business customers, because an integral part of the application included end-to-end process responsibilities.

We worked out a plan for my helping her to convince her senior leadership team of the benefits of applying for the MBA. Typically, the rate of improvement in a unit in the year they apply for the MBA doubles over other years as the enthusiasm of competing for the national quality award spreads throughout all the people in the unit.

I also agreed to help convince the head of the network services division to participate with her unit. Our network services division is responsible for the operation of our network and they had been doing a great job on improving our network reliability.

I agreed to advise and support Jane and her application team. Before the end of 1992 I had two MBAs "in the bank" and a commitment by my two largest units to apply in 1993. This was a pretty good year for a quality vice president.

AT&T SUCCESSES AND SHORE SETBACKS

In 1992, the quality of the New Jersey shore water worsened and the quality of AT&T operations significantly improved, a reversal from 1991. However the annual results turned out, I was committed to continuous improvement of the management system's capability to produce better results in future years at both the New Jersey shore and at AT&T.

They Must Know How Much You Care

Before the Baldrige Award ceremony, one of my associates asked if I was interested in walking over to the Vietnam War Memorial with her. I had thought it looked like a large tombstone from pictures on television, but I reluctantly went along to see it in person that morning.

Having to walk by 59,600 names of the Americans who died in Vietnam made me really appreciate how lucky I was. The memorial up close has a different impact than the photographs. It was larger than I had realized and I felt very small and unimportant compared to all those on the wall.

I had been assigned to Nha Trang, Vietnam in January, 1968 as a young second lieutenant in the U.S. Army. During the first three months, I was assigned as the night duty officer in charge of the Nha Trang communications center for the Strategic Command Center. As a result of working nights, I was asleep in the junior officers' hotel in Nha Trang the morning the Tet Offensive started. Nha Trang was the first city hit, and we were all caught by surprise.

While I was sleeping that morning, a company of Viet Cong came into town and made one of their first stops at my hotel. The old South Vietnamese guard at my gate told them no one was in this hotel, so they went on. The guard then woke me up and told me the city was under attack. He left the area and I was left alone in the hotel as gunfire broke out in the surrounding area.

That morning the Viet Cong killed 26 senior officers and their support staff in the hotel down the block from the hotel where I had slept

Fortunately, I was left alone for the two days of fighting. I was a junior officer assigned by myself to an inferior hotel, and as a result did not come directly under attack.

I could have been a name on the Vietnam Memorial wall. Instead, I was going to a celebration with my friends, to be recognized for AT&T's quality work by the president of the United States.

I asked park service personnel and veterans groups that day at the wall how many of our South Korean allies, or our allies from other nations such as Australia, were killed in Vietnam. No one knew. I asked how many of our South Vietnamese allies were killed. No one knew. I asked how many Vietnamese civilians had been killed. No one knew.

I was referred to several books that might include such information. Over the next few years, I bought a dozen books on the Vietnam war but none included the data on all the Vietnam conflict casualties. I searched the Internet but found nothing. Then, in 1995 I spent a day in the Tufts University library in Medford, Massachusetts, going through the stacks.

continued

I found a fact-filled book by Guenter Lewy, a professor from the University of Massachusetts, titled *America in Vietnam*. The information I wanted had been researched and documented, but not yet referenced in the computerized library information systems.

The best estimate of casualties in the South is 582,000, of which half (300,000) were South Vietnamese civilians. Therefore, the U.S. military casualties of 59,600 was about 10 percent of the total losses in South Vietnam. The best estimate of casualties in the North is 731,000 of which less than 10 percent (65,000) were North Vietnamese civilians and 90 percent (666,000) were Viet Cong or North Vietnamese Army. The U.S. casualties in the Vietnam conflict accounted for less than 5 percent of the total casualties (1,313,000) from 1964 to 1975. Of the total casualties inflicted on both sides of this war the numbers were close, with 55 percent of the losses in the north and 45 percent of the losses in the south. However, when the U.S. pulled out on March 29, 1975, we left our South Vietnamese allies outnumbered 2 million troops to 1 million troops, leading to a North Vietnamese military takeover of the South, in spite of the peace treaty that was signed before we pulled out. I could find no records of the South Vietnamese losses after we pulled out.

When I found out how few knew the total casualties in Vietnam, and how hard it was to find out, I realized we really didn't care enough about the people we were trying to help and protect. To most, Vietnam was not about helping people, it was about Democracy vs. Communism. I learned if you do not care, really care, about the people you are trying to help then you cannot really help them. A good teacher once told me that before people care how much you know, they must know how much you care.

Caring helps you go all the way when faced with a setback. We had put our boots on, and Tet was a setback. Because we didn't care enough, we didn't go all the way. Because the North Vietnamese cared enough about their country, they kept their boots on, they went all the way.

Before making a decision to help, be sure you care. Before you put your boots on, be prepared to go all the way. Recognize that there will be setbacks; be prepared to overcome them to reach your goal.

Lessons Learned (1992)

1. Setbacks will happen, keep your boots on when they do.
2. Those controlling the resources must be informed and supportive of the improvement efforts.
3. National goals and local solutions work better than no goals and national solutions.
4. More people working on a solution means more solutions.
5. Sharing local solutions (knowledge transfer) is a useful role for support staffs.
6. They must know how much you care, before they care how much you know.

CHAPTER SIX

The Dolphins Are Back!
(1993)

AT&T UNITS READY FOR QUALITY ASSESSMENTS

Two of the largest parts of AT&T, the consumer long-distance marketing unit and our network operations unit, had made significant improvements over the past few years and were ready to undergo evaluation of their quality system by the national Baldrige quality assessment process. A Baldrige application has to be submitted by April, so early in the year I worked with Jane Redfern and the team to prepare the units and the application. The application required 75 pages of answers to the Baldrige criteria questions on how we managed the business. Preparing the units meant working with them on needed business improvements if they wanted to reach the finals in the MBA process, which meant getting a site visit. The business units worked very hard and submitted their first MBA application in April 1993. But their hard work didn't stop in April. It continued throughout the year supporting the business process improvements they knew were still needed.

In our AT&T manufacturing sector, AT&T Microelectronics Power Systems had performed very well in our AT&T CQA assessment process and had also won the U.S. Shingo Prize for manufacturing excellence in 1992. I felt they would be a good candidate in 1993 for the Baldrige Award in the manufacturing sector. However, the business unit head had set his goal to be better than his competition from Japan. Therefore, he had a plan to apply for Japan's national quality award, the prestigious Deming Prize, which had been opened up to worldwide competition several years earlier. However, it had never been won by a U.S.-based manufacturing company.

I advised the unit head that the Japanese assessment criteria were more prescriptive than U.S. national quality award criteria and that Americans might react negatively to taking away their freedom of choice on the improvement methodology. He assured me that he was aware of this issue and he had assurances from the Japanese that they would be sensitive to U.S. cultural issues, while still giving them a rigorous evaluation using the Japanese award criteria.

I decided to convince our New Jersey-based AT&T Capital Corporation, a leasing and financing unit, to apply for the new New Jersey Quality Achievement Award, which was based on the national Baldrige criteria. In 1993, the governor of New Jersey signed an executive order establishing a New Jersey state quality award, to be administered by Quality New Jersey using a volunteer team of examiners. Our major concern at QNJ was finding enough New Jersey-based companies who would be qualified applicants, ones that had developed a quality management system that could meet the demanding Baldrige criteria. The New Jersey-based AT&T Capital Corporation financial services unit had been assessed as part of our 1991 and 1992 AT&T Chairman's Quality Award, using the same Baldrige criteria, and they had been hard at work for a few years improving their operations based on the feedback received in their CQA assessments. While Capital Corporation wasn't sure they were ready to apply for a state award, I encouraged their quality director, Kirk Perry, to apply and AT&T Capital Corporation decided that an external assessment of their progress might be useful for further improvement.

SHORE QUALITY IMPROVEMENT PLAN

The 1992 disruption and setback on our shore improvement effort had our QNJ environment team concerned that the shore municipalities might have lost faith in the support they could expect from the state toward our goal of zero beach closures.

At our first QNJ environment team meeting of the year the team decided to continue our improvement support efforts, which included the following steps:

- Establish goals and measures for continuous improvement of the Jersey shore water quality.
- Conduct root cause analysis of the shore water quality problems.
- Develop countermeasures for the root causes and share them with shore municipalities.
- Assess and recognize excellent municipalities and counties using our QNJ Shore Water Quality Award.

We also decided a few new areas also needed improvement:

- Improve our sharing of best practices with the shore municipalities.
- Improve our partnership relations with both the executive and legislative branches of state government.
- Improve our partnering with other New Jersey environmental groups concerned with the Jersey shore.

From 1988 to 1992 we had gone from over 8 beach-blocks closed per day of the summer season due to pollution (855 beach-block-days of closings in the 100 summer day season of 1988) to about 1 beach-block closed per day due to pollution (111 beach-block-days of closings in the 100 summer day season of 1992). In spite of the 1992 setback, we decided to maintain our goal of zero beach-block-day closings by 1996. We held to our goal because we believed that the improvement programs that had been put in place could support continued improvement results by the shore municipalities by addressing the sources of pollution.

Floatables

In 1989 a Floatables Action Plan was developed by the New York–New Jersey regional office of the Environmental Protection Agency (EPA), aimed at capturing floatables upstream. Ninety to ninety-five percent of the floatables were wood (old piers and broken trees from storms). The Floatable Action Plan called for collecting debris in the New York–New Jersey harbor so that it would not float down to the Jersey shore and pollute our beaches. The debris collected was loaded by the Army Corps of Engineers into garbage scows provided by the New York City Department of Sanitation and disposed of at the Fresh Kills landfill on Staten Island, New York.

The amount of floatables collected in the New York–New Jersey harbor since the program was initiated is as follows:

- 1989—545 tons
- 1990—795 tons
- 1991—701 tons
- 1992—958 tons

The New Jersey DEP also operated "Operation Clean Shores" program that used inmates to remove debris found along New Jersey rivers, bays, and shorelines. In 1991 New Jersey had 4,688 tons of debris picked up by our inmate program and in 1992 that rose to 5,030 tons. A significant part of this was picked up along the rivers, bays, and shores north of the beach areas to prevent it from becoming floatable debris. Dave Rosenblatt's staff ran this program.

Sewage

The most important improvement program was aimed at the sewage system and supported by the New Jersey Sewage Infrastructure Improvement Act. This act, proposed by the New Jersey DEP, provided state funding for information system mapping of municipal sewage and storm-drain systems.

The act applied to the four shore counties and to the 90 municipalities along the Jersey shore. In 1993, $5 million was made available from the state for use by the municipalities to start the program and another $6 million was planned for 1994. The act also required shore municipalities to monitor their storm-drain water and make improvements to their infrastructure when measures indicated a problem.

Taking measures in the ocean or bays near swimming beaches tells you about a problem at the output of the process. We also wanted to move our measures upstream (that is, up drain) in the process to find out about problems in the process before we had problems in the output. In this way it would be possible to fix the problem before the ocean or bay became polluted. In business, these upstream measures are called process measures, which are managed to ensure that desired results objectives (outputs) are achieved.

Street Litter

A member of the New York–New Jersey regional EPA office, Peter Brandt, had been on our team since 1990. In 1992, the New York–New Jersey Regional EPA office initiated a new program called "Clean Streets–Clean Beaches." This was a campaign aimed at educating school children, and other citizens, that street trash ends up on our bay and ocean shores via our storm-drain system. This was another preventive program being supported by our government, this time an education one.

Five northern counties in New Jersey had also implemented "Clean Streets–Clean Waterways" and "Clean Streets–Clean Streams" programs prior to the 1992 EPA "Clean Streets" program. These programs were aimed at prevention of litter. Prevention is much more efficient and effective than litter clean-up programs. Of course, if we don't prevent litter then we must clean it up.

Beach Debris

One of the programs approved by the legislature and implemented by the DEP in 1993 was an "Adopt a Beach" program. Under this program the DEP encouraged and recognized volunteers for adopting a stretch of beach, which the volunteers would clean up regularly by picking up any debris or trash. This wasn't punishment for minimal performance standards, it was recognition for excellence. The program used volunteers and therefore had a minimal cost. Dave Rosenblatt's staff also ran this program. I think the state was beginning to understand how a quality approach could work.

SUPPORT OF A QUALITY APPROACH

The state was now supporting information systems development, process management, process improvements, and volunteer recognition. The New York–New Jersey area regional EPA office was working with both New Jersey and New York City, as a team, to prevent New York–New Jersey harbor debris from floating down to the Jersey shore. In addition, the EPA was supporting education as a means to prevent litter and trash from moving from streets to beaches.

The headlines after the Memorial Day opening of the 1993 summer season at the shore read: "Jersey beaches and water quality pass." A passing grade from the ever-critical press felt pretty good. The article reported that "the shore had excellent water quality, beautiful weather, and only one beach closing that day."

We had a passing grade, but had not yet made the Dean's list.

That one closing was from a clogged grease drain in a Cape May hotel. Cape May usually had excellent water quality and this problem was a particular disappointment to one of our AT&T team members who happened to be the daughter of the mayor of Cape May. She worked for AT&T and because she had an obvious interest in the shore, she had joined our team the prior year.

Even though we did have one beach closing on opening day, the sun was in the sky, the water was clear and it was beginning to look a lot like a quality New Jersey at the shore. Our Quality New Jersey environment team was feeling pretty good. We had overcome our 1992 setback by educating both the executive (management) branch and legislative (funding) branch of state government on the value of continuing our quality approach to shore water quality improvement. We approached continued government support for our improvement efforts as just another quality problem, which has root causes and potential countermeasures. A quality approach could be applied to the government management and funding problems, as well as the sewage and litter problems.

One of the root causes of our losing the 1992 DEP ocean monitoring program funding was that the legislative branch of state government was not informed and educated about the value of the quality approach we had been using to clean up the Jersey shore. The initial countermeasures we implemented were to ensure that key state administrative executives and state elected representatives were informed and educated about the value of the quality approach we had initiated, to avoid another disruption of our efforts to monitor and improve the shore water quality.

Our 1992 letters to key state legislatures and the DEP administration, with data on the results we had achieved, resulted in assurances of continued state support by both the legislative and executive branches of government for our Quality New Jersey approach to continued improvement of the Jersey shore water quality. We could begin to see signs of support for a quality approach by the government organizations involved in the New Jersey shore water quality improvement efforts. Blame, turf, emotion, regulations, and fines were beginning to be replaced with teamwork, process, facts, goals, results, and recognition. I like to think our little team was leading and supporting this change. It was the government organizations that were making the change and improving the New Jersey shore water quality.

COMPARE AND COMMUNICATE THE RESULTS

The communications of results achieved, to our New Jersey representatives at the state and federal level, included a comparison of our results to other states during the same period, to demonstrate the value of a quality approach.

The only source of data to compare the water quality results and trends from state to state is the National Resources Defense Council (NRDC) annual reports that are compiled by the NRDC by calling individual state DEP agencies for data on their state results. While this data is incomplete, because the EPA has no national requirements for monitoring our U.S. shore water quality, it can be used to make some comparisons on improvement trends from state to state.

One of the problems with comparing the absolute results on a state-to-state basis is that only a few states monitor the water quality of their entire shoreline. New Jersey does monitor the entire 127 miles of its ocean shoreline as well as the bay shoreline, but in 1992, California only monitored 144 miles of its much longer shoreline and Florida only monitored 97 miles of its much longer shoreline. New York monitored 56 miles of its shoreline.

Another problem is that not all states actually close a beach when the water quality does not meet the recommended health safety standard. Some states simply issue an advisory and leave the beach open to the public to use "at their own risk." In 1992, New Jersey was the only state in the nation that both monitored its entire shoreline water quality and closed beaches when the recommended standard was not met. All states should do both to protect the health of swimmers and to provide shore towns, who rely on tourism for a livelihood, with a strong motivation for meeting the minimum water quality standards. The EPA should make this a national requirement of each state to protect our national shoreline.

The numbers of closings and advisories in New York, Florida and California had each increased by over 300 percent during the same period that New Jersey had made its major improvement. New York, California, and Florida all had been below 200 beach-block-days of closings in 1988 and 1989 and all three increased to above 600 in 1992, an increase of over 300 percent. During the same period, New Jersey had decreased from 855 beach-block-days of closings in 1988 to 111 in 1992, a decrease of more than 85 percent.

To do a fair comparison, one should compare the "closings and advisories" that occurred per actual shore mile that was monitored, recognizing that a state could have increased the amount of shore-line monitored and that could have contributed to the increase in closings or advisories. The water quality standard is not met when there is a closing or an advisory.

When the results in these states were compared on a per-mile-monitored basis, it also showed that New Jersey had a major improvement and the three other states had all become significantly worse. New Jersey had been the worst in the nation in 1988 on both the total number of closings and on a closings-per-mile-monitored basis. In 1992, New York took over the undesirable position as leader on both the total number of beach closings and advisories as well as on a closings-per-mile-monitored basis.

California had the worst trend on water quality with an increase in their closings and advisories per mile monitored from 0.4 in 1989 to 4.2 in 1992—an increase of 1050 percent in just three years. New York increased from 4.9 in 1989 to 14.3 in 1992—an increase of 292 percent. Florida had increased from 2.6 in 1989 to 5.6 in 1992—an increase of 215 percent. New Jersey, using our quality approach, had decreased from 6.8 in 1988 to 0.9 in 1992—a reduction of 87 percent.

These numbers provided a strong indication that we were doing something right in New Jersey that was not being done in other states. We hoped this data would help inform and educate our New Jersey government leaders to support the continued funding of the quality efforts that had resulted in the shore water quality improvement in New Jersey. I communicated these results and comparisons to help avoid another disruption of our shore improvement efforts. Communicating results was one of our countermeasures to the funding problem.

BAY POLLUTION OF SHELLFISH

We had a significant portion of our New Jersey bays not only closed to swimming, but also closed to the shellfish industry. In 1992, we had about 20 percent of our New Jersey bays closed to shell fishing due to pollution.

The June 1993 Natural Resources Defense Council (NRDC) report stated that 40 percent of New Jersey's shellfish waters were harvest-limited due to pollution. This compared with 81 percent of shellfish beds in Florida and 90 percent of California's shellfish beds which were classified as harvest-limited due to pollution.

Sewage leaking into storm-drain pipes, or being directly dumped untreated into the bays, were also indicated as the primary causes of problems in other states. An inadequate sewage infrastructure, which was also inadequately maintained, was polluting our shores from north to south and from east to west—"from sea to shining sea." Perhaps it was the pollution that was giving our sea its shine.

It was becoming clear that we had made significant progress on our New Jersey ocean beach water quality, which now seemed to meet ocean beachgoer expectations. However, most people in New Jersey did not even have expectations for improvement on the bay side. People had felt that the bays were to be left to boating only, given their level of pollution. Many of our valuable New Jersey bays had fallen into a category with the Hudson River—swimming was not expected and you better not fall off your boat into "that" water.

Our QNJ environment team believed we could maintain New Jersey's ocean improvement and continue to improve our bay water quality. We also believed we could begin recovery of our bays. One day we would be able to reopen all our Jersey bays to the shellfish industry and to swimming.

The idea that we could recover our bays was not just a dream, it was based on the good work that we had seen in bay shore towns like Stafford Township, which had won our 1992 QNJ shore water quality award by taking the storm-drain pipes out of the bay. Stafford Township had implemented an innovative idea. If you can't take all the pollution out of the pipes, take the pipes out of the bay.

With the deployment of the ideas we had already identified to reduce or eliminate bay pollution, along with more shore town innovative ideas, we could make our New Jersey bays safe for swimming, and livable for shellfish.

DEDICATED FUNDING SOURCES

In spite of our education of the legislative and executive branches of state government, we still were concerned with funding support for our programs, recognizing the changes in leadership that may occur. We needed a dedicated source of funds from people who really cared about the shore.

In 1993, a New Jersey license plate with the motto "Shore to Please," along with a design of a lighthouse on the beach, was offered through the Department of Motor Vehicles (DMV).[1] This "designer plate" could be purchased for a $50 extra initial fee and $10 per year extra annual fee. These voluntary extra fees were then dedicated to funding shore improvement projects under the control of Dave Rosenblatt at the New Jersey DEP. The projects funded included such things as the Operation Clean Shores and the Adopt a Beach programs. Residents' purchase of these optional plates provided a cushion of funds that protected these shore clean-up programs from further legislative budget cuts. It didn't take long before we had 60,000 people proudly using these shore designer plates, which provided about $1 million per year toward our clean-up efforts.

I added the additional cost of a "vanity plate" to the cost of my shore "designer plate" to become the proud owner of a shore license plate with the initials "QNJ."

THE IMPORTANCE OF STANDARDS
AND MEASUREMENTS

In line with our work on the New Jersey shore, in July 1993, our New Jersey representatives in Congress proposed a bill to require uniform national ocean water-quality testing standards. However, the EPA failed to endorse the bill, saying that they already had sufficient authority to establish national beach water-quality standards—if they wanted to. The EPA has been saying they have this power since 1988, but they have never used the authority. Without EPA support, and with some representatives from other states arguing that mandatory testing of our ocean water would be too expensive, the proposed law to measure our ocean water quality was defeated. (Through June of 1997, no national law had been passed to require testing of our shore water for pollution, and the EPA has failed to use the authority they say they have to issue a regulation requiring testing of shore water.)

The legislators who voted against the bill to measure our ocean water quality do not understand that a quality approach requires measures and standards. Measures are needed to set goals for excellence and to monitor progress against those goals. Measures and data are needed to analyze root causes and to prioritize appropriate countermeasures. Without measures, teams or companies or government can't develop appropriate countermeasures. Without measures of our ocean water quality we can't even know if people, or fish, have healthy water to swim in.

There is a small cost to measuring, but measurement costs can be kept to a small part of any improvement effort using statistics taught in high schools. And a quality improvement project based on the measures can have a great payback. For example, before we began improving the New Jersey shore water quality, shore tourism business was declining. After our shore water-quality improvements, the New Jersey shore tourism business became the fastest growing market sector in New Jersey, increasing by about $1 billion per year.

With New Jersey's 6 percent sales tax, and an income tax that averages 3 percent, the state's increased revenues from the shore tourism increase is more than $60 million per year, each and every year. Over five years of such growth, the state government would add over $300 million per year in new revenue without any new or increased taxes. The power of an improved economy is that the government gets a percent of the improvement. Our project was producing an improved state economy, and an improved state environment—at the same time.

The cost of the annual measurement of the shore water quality was about $1 million per year, including the four shore counties' measurement costs and the state-level analysis. The state was getting back hundreds of times its investment from the improved economy on the shore. In addition, the beach-going voters in New Jersey were happy—and the dolphins were healthy and back in increasing numbers (Photo 6-1).

People in government, industry and academics tend to argue that testing is too expensive because they do not understand how to use simple statistics and sampling procedures. Every day throughout industry, small numbers of measures, combined with the use of statistics, provide adequate quality control capability with known confidence levels.

The U.S. Census Bureau has a strong capability in the use of small numbers of measures and statistics to measure all kinds of things about the country. In fact, one of the most famous U.S. quality gurus worked at the U.S. Census Bureau, Ed Deming. This is the same Deming for whom the Japanese named their national quality prize. In 1997, when Congress proposed in the balanced budget bill that the next census not use sampling, it demonstrated

Photo 6-1. Dolphins Playing in the Ocean

its lack of understanding. The president vetoed this proposal, which would have just wasted money by increasing measurement costs unnecessarily.

Measures, plus data derived from sampling, provide the basis for a quality approach to improvement. Without data, a quality approach to improvement is not possible. Absent data, people react emotionally and instead of being able to reach consensus based on data, we are left to argue our own opinions, never knowing which opinions data would have shown to be incorrect. While facts alone do not produce quality improvement, they are a necessary ingredient. It is too bad our federal government did not yet understand this basic quality point. It didn't look like a quality approach was understood at our federal level of government in 1993.

However, Vice President Al Gore had just announced the Reinventing Government Initiative, based on a quality approach to improve federal government agencies. I was subsequently requested to spend a few days advising one of the federal agencies, the U.S. Customs Agency, on their approach to reinventing Customs. The first review was disappointing and I questioned the value of spending a second day with a group that seemed convinced they were restrained from changing by 400 laws, some of which were over 200 years old. Happily, I can report that this group didn't quit, they kept their boots on, and developed a great proposal for reinventing Customs. In fact, they were the first agency to receive congressional approval for their reinventing government plan. Of particular help was the fact that the Customs Agency commissioner was a former congressman who knew how to get Congress to give him relief from the 400 old laws that affected what Customs had to do.

My experience has shown that we need the application of a quality approach at every level of government, and in all branches of government. In addition, we need business–government teams that address the responsibility and accountability of all four levels (federal, state, county and municipal) of government on key processes, such as those that affect the environment, education, and health care. Our QNJ environment team is just one of many such teams required in each state to help improve our quality of life.

CONFRONTATION OR PARTNERSHIP?

One of our 1993 team objectives was to partner with other New Jersey environment groups interested in the shore water quality. One group we contacted, Clean Ocean Action, requested that we partner with them and several other environment groups, in a lawsuit to prevent the EPA and the Port Authority of New York and New Jersey from issuing permits for moving New York–New Jersey harbor sediment out to sea. They first wanted the harbor sediment tested to see if it met EPA standards, which required ocean sediment to be essentially toxin free. Most people suspected the harbor sediment was not toxin free, but the EPA was again not measuring. The other environment groups believed that small amounts of toxin in the harbor sediment might accumulate in marine life.

Our team decided we did not want to partner on a lawsuit on this issue for two reasons. First, our team objective was to achieve zero pollution on the Jersey shore as measured by tests of our beach water quality. Our data showed moving harbor sediment into the ocean was not a source of New Jersey beach water-quality problems. Since it was not a source of the New Jersey shore water-quality problem we did not need a countermeasure, such as the planned lawsuit. The other environmental associations had a different goal than we did. Second, our team approach to solving problems was through partnerships with government using a quality approach. Confrontation with government using a legal approach was not compatible with our quality approach. You can't ask to sit at the same table to jointly analyze data and develop root causes and solutions in the morning with someone and take them to court in the afternoon.

While I do not object to lawsuits or confrontations, which are occasionally needed to get action, I believe a team has to pick one approach or the other—confrontation or partnership. Our QNJ environment team, and the entire QNJ organization, picked the partnership approach.

Had our team had the goal of preventing any level of toxin from being moved from a river to the ocean, we would have considered possible prevention approaches prior to a court action.

Since sediment moves down a river to a harbor and then needs to be moved out to sea, the river sediment should be kept toxin free. First, all possible sources of river pollution from toxins should be prevented and cleaned up. Second, the river should be managed to minimize the sediment it carries by allowing the river to widen when it floods. This would reduce the sediment that accumulates in the river harbor. Both of these are upstream preventive measures that a partnership effort could pursue.

While we did not partner in this lawsuit, we did meet with some of the other environment groups to explain our quality approach, in case they chose to use it instead.

After three years of no dredging in the New York–New Jersey harbor due to the environment association lawsuit, sediment build-up got so high that it limited deep-hull tankers from entering the harbor. The harbor began losing business, and jobs, as ships went to other harbors. The sediment dispute was putting a $20 billion per year harbor and shipping business, and the associated New York–New Jersey jobs, at risk.

While three years were spent on confrontation in the courts, the root causes of the sediment in the harbor, or why we had toxins in the sediment, were not being addressed. The root causes included such actions as elimination of wetlands that would have allowed the Hudson River and the sediment it carries to spread out upstream. Instead, the river and the sediment rush to the harbor, carrying tons and tons of sediment to the harbor mouth. Another root cause was an abandoned factory (the Diamond Alkali site) on the river that had once manufactured Agent Orange for use in Vietnam. Alkali metals are a group of highly reactive metallic elements including lithium, sodium, potassium, rubidium, cesium, and francium. While some will blame industry for the abandoned factory, it should also be remembered that it was our federal government that was the customer for the Agent Orange toxin. I felt lucky to escape Agent Orange when I was in Vietnam, where it was sprayed on both vegetation and troops. Twenty-five years later, it shows up again in my home state's waterways.

The abandoned Agent Orange factory site and other sources of toxins should have been identified and cleaned up long ago. In 1996, the New York–New Jersey Regional EPA office had just com-

Just before the opening of the 1996 summer shore season we had an oil spill off the coast of New Jersey, as a deep hull tanker attempted to transfer oil to a smaller ship in the middle of the night while at sea. The ocean turbulence caused the two ships to pull apart and spill thousands of gallons of oil that washed up on the New Jersey shore during the opening Memorial Day weekend. This accident would not have happened if the ship had been in harbor tied up at a dock. Not being able to bring tankers into harbor to unload their oil increases the risk of oil spills. The environment association lawsuit was not originally about the Jersey shore water quality, but it did affect it.

Within a few months of that 1996 opening day oil spill, the Clean Ocean Action environment group dropped their three-year lawsuit against the EPA and agreed to support an EPA and Army Corps of Engineers decision to restart the New York–New Jersey harbor sediment dredging with continued sediment dumping offshore. Also, alternative disposal methods would be studied while the sediment with high toxin levels would only be dumped in the ocean for one more year. This agreement seems like one that should have been reached three years earlier without the expense and time of a lawsuit, without putting at risk the harbor jobs, and without putting at risk the New Jersey shore water quality to offshore oil spill accidents.

pleted a $12 million study of the Passaic River within three miles on each side of the abandoned Agent Orange site and began working on a cleanup plan in 1997. Unfortunately, because this cleanup was not addressed earlier, people have fished for crabs in this area for years—and some still are, in spite of signs posted in five languages warning of the danger.

I CAN SEE MY TOES— AND THE DOLPHINS ARE BACK

While a quality approach was not used to address the New York–New Jersey harbor sediment problem, the quality approach that we had been using on the New Jersey shore water quality problems was paying off for us in the summer of 1993.

In August of 1993, Lester Jargowsky, Monmouth County's Health Officer and winner of our 1991 QNJ Shore Quality Award, was prompted to send a telegram message alerting shore police departments to shore problems being caused by natural phenomena. (A problem being caused by the clean water.)

"The water's too clean," he said jokingly. "It's unbelievable. This is the first time ever we've had good water visibility to depths of 20 feet." People were seeing fish in the ocean—striped bass, jellyfish, squid, flukes, sea robins, and even some tropical fish. "In the past, there never was more than six inches of visibility in the water." Jargowsky added, "There is nothing to fear. The fish are just part of the shore experience like the clams, gulls, and sea breezes." Swimmers could not only begin to see fish again, they also could see their toes.

In 1993, more than fish returned to our shores. A few mammals also returned to the New Jersey shores. Half of the 5,000 North Atlantic bottlenose dolphin population had died of pneumonia in 1988. With the clear water in 1993, the dolphins were back in full force!

With the dolphins came tourists. Island State Park, a state park on the coast, reported a 23 percent increase in shore visitors during the summer of 1993, compared to the summer of 1992.

My kids also returned to the Jersey shore in 1993. John, my youngest, was now 16. He returned home from a day at the Jersey shore with his buddies and ran in the house yelling: "The water was beautiful! Dad, you should have been there—we had a ball!" When I heard those words, I knew that was what I had been working on the shore to achieve.

1993 QNJ SHORE QUALITY AWARD WINNERS

In 1992, Avalon was charged $1,807,763 by the Cape May County Municipal Utilities Authority (MUA) to treat 287 million gallons of effluent (sewage) outflow, but Avalon only pumped 159 million gallons of effluent (sewage) inflow. This meant 128 million gallons of storm, or ocean water, entered the sewage system by inflow, or infiltration, costing Avalon home owners in excess of $600,000 per year for sewage treatment costs of storm and ocean water.

When they analyzed this problem, Avalon found that 64 million gallons of inflow water entered through open sewer units. As a countermeasure they capped the open sewer units. They found that 42 million gallons had infiltrated through sewer manholes. As a countermeasure they sealed the manholes. They also found that

22 million gallons infiltrated through the old leaking sewer mains. As a countermeasure they replaced the old sewer mains.

The old sewer system leaked both ways, of course—storm water into the sewer main and sewage into the storm-water pipe. Leakage of storm water into the sewage system resulted in higher bills for sewage treatment costs. Leakage of sewage into the storm-drain system resulted in pollution of our waterways. Avalon not only improved their ocean water quality, they found that by reducing the effluent in Avalon they could increase the "affluent" in Avalon.

Nineteen ninety-three was the third year for our annual New Jersey Shore Quality Awards and the first year for a repeat winner. The municipality of Avalon, which had won in 1991, also won again in 1993. They had maintained perfect ocean water quality and continued their improvements in the storm-water and sewage system. They had focused on repairs, capping sewer units, sealing manhole covers, and installing tide flex-valves in the catch basin at the end of the pipe. Avalon had found that maintaining a sewer system that didn't leak into the storm-drain system kept the ocean water clean. They also found that maintaining a sewer system that didn't let storm or ocean water leak into the sewage pipe saved money.

Avalon was learning how quality improvement can improve output and reduce costs. Their quality approach was also increasing tourism revenue from very happy beachgoers. Avalon's winter population of less than 3,000 rose to over 30,000 during the summer months. The Avalon quality approach was reducing pollution, reducing costs, increasing revenue and meeting beachgoer expectations.

QUALITY ASSESSMENTS

With things going so well at the shore in 1993, I decided to submit an application from our Quality New Jersey environment focus group for the Rochester Institute of Technology–*USA Today* Quality Cup Award. *USA Today* awards this prize annually to recognize successful quality teams. I had hoped recognition of our team might increase support throughout New Jersey for our partnership approach, and increase awareness for replication in other states.

Since I believed in using the plan-do-check-act cycle, not just advising others on how to use it, this was also an opportunity to have a free check done on our own QNJ environment team. Our team accomplishments included a factor of 10 improvement in the New Jersey shore water quality and a 60 percent increase in New Jersey shore tourism revenue over a five-year period. We had formed a citizen–business–government partnership that produced the most improved shore water quality in the nation, plus the dolphins were back and healthy.

The Quality Cup Award covered five key areas:

1. Process Improved (shore pollution prevention processes)
2. Quality Team Approach (plan-do-check-act)
3. Results Achieved (90 percent pollution decrease, 60 percent tourism increase)
4. Measures of Change (beach-block-days of closed beaches, 855 to 88)
5. Benefits to Customers (beachgoer satisfaction, shore jobs, dolphins back)

Although we didn't win the award, we found that preparing the application provided a simple five-page overview of what we had done in a form every QNJ focus group could understand and use. We shared the Quality Cup application format along with our own completed application (see Appendix D) with the several other QNJ quality teams that were working in other sectors to improve the quality of life in New Jersey, which included education, health care, government, business manufacturing, and business service. The Quality Cup format, and our environment team example application, proved to be a benefit to all QNJ teams by giving a simple and standard way to check QNJ quality teams.

AT&T and the 1993 Quality Awards

AT&T Consumer Communications Services' network services division got the news that they didn't win the Malcolm Baldrige National Quality Award in 1993. After a week-long examination, the Baldrige site-visit examiners' report cited good results. However, further areas for improvement were identified that were required to reach the excellent level expected of national award winners. These identified areas for improvement became part of our business improvement plan for 1994.

The credit division of the AT&T Capital Corporation had applied for the New Jersey Quality Achievement Award and they received the good news in 1993 that they had become the very first winner of the award. The Quality New Jersey state award had been developed to motivate business and government—and in 1996, education and health care—to continuously improve, not to win awards, but to make New Jersey the number one state in which to live and work

QNJ QUALITY CONFERENCE

A very happy Gerri Gold, president of the New Jersey-based AT&T credit unit, accepted the first New Jersey Quality Achievement Award from the New Jersey commerce commissioner, Barbara McConnel, at our 1993 QNJ quality conference. The DEP commissioner, the commerce commissioner, and I presented the QNJ Shore Quality Awards to our 1993 winners, Avalon and Cape May County. The commissioners were beaming that day, with the smile of proud parents when they see their children graduate from school. Recognition is more fun than fines.

One of our guest speakers at that year's conference was Curt Reimann, director of the Malcolm Baldrige National Quality Award program with the U.S. Department of Commerce, National Institute of Standards and Technology. Curt came to support our initial state quality award, which was based on the Baldrige criteria. Curt also heard about the work being done by our shore towns, following a quality approach, and left very impressed with the work of our QNJ focus groups. Subsequently, I was invited to be a guest speaker at the February 1994 "Quest for Excellence" conference in Washington, D.C. to share my QNJ environment focus group story and results. The invitation to speak with the national quality award winners made me feel like one of the winners that year.

Lessons Learned (1993)

1. Recognition, combined with financial support, can accelerate replication of best practices and the rate of improvement.
2. Punishment does not accelerate improvement.
3. After solving a large problem, you're likely to discover a smaller problem.
4. A visible improvement is required to satisfy customers.
5. Quality improvements should also provide cost improvements.
6. Partners require a common goal and a common approach.
7. Partners don't sue each other.
8. You need to stay focused on your goal and on eliminating the sources of the problems that affect reaching your goal.
9. Without measures you can't develop appropriate countermeasures.
10. Recognition is more fun than fines.

CHAPTER SEVEN

Another Budget Cut
(1994)

UNDERSTANDING THE
COST–BENEFIT RELATIONSHIP

In January of 1994, a new governor, Christine Whitman, took office in New Jersey. Whitman, a Republican, defeated Democratic governor Jim Florio with the promise of reducing state income taxes by as much as 30 percent.

Two years earlier, in 1992, the shore improvement effort had been set back by the new Republican assembly's budget cut that reduced the state sales tax from 7 percent to 6 percent. While only a reduction of 1 percent in the prices charged consumers, it resulted in a 14 percent reduction in the revenue received from the sales tax. The 1992 budget cuts resulted in a shutdown of the state DEP's shore-water monitoring program until we had another shore pollution problem a few months later. This program provided the measures for our shore water quality trends and the information needed to know where problems occurred, to drive fast repair. Eliminating this measure, along with the shutdown of the DEP coastal monitoring team, resulted in a setback to our efforts to continuously improve our Jersey shore during the summer of 1992.

In response to the shore pollution problems of 1992, the DEP shore monitoring budget was restored, but only after the 1992 summer shore season was over. The restored DEP monitoring program, combined with our shore improvement programs, enabled the shore to have its best year ever in 1993.

The state budget in 1993 was helped by budget cuts in some areas (reducing outflow of cash) and by the increased shore tourism revenue (increasing inflow of cash). Like a successful business, a successful state needs to manage both its expenses and its revenues. In 1993, the sales tax stayed at 6 percent and we had a great shore to enjoy, producing increasing tourism revenues.

Early in 1994, in order to follow through on her promise of reducing the income tax, our new governor put a hold on spending for a number of projects. One of the projects included in the 1994 across-the-board spending hold was the Sewage Infrastructure Improvement Act grants, which provided some funds to the shore municipalities for their shore improvement efforts.

Now, just two years later, another budget cut was affecting our shore improvement efforts and, potentially, the growing shore tourism revenue stream. Although the shore monitoring program was restored in 1992, we had not fixed the government shore funding support process to prevent reoccurrence of budget cuts to our shore improvement programs. The enormous benefits of the shore monitoring and improvement programs to the state, by improving both shore water quality and state revenue, had not yet been recognized by the new governor.

The budget was again being cut by new government officials who had been elected on a promise to reduce state taxes. I remembered that our voter polls always had cutting taxes as the number one priority. As a voter and taxpayer, I agreed with that priority. However, I had come to realize that the small investment required to maintain our shore quality also protected and increased our shore tourism revenue, producing a 30-to-1 payback to the state. The state benefited from both a 6 percent sales tax and an income tax of about 3 percent on the $15 billion per year shore revenue. The sales tax alone accounted for $900 million per year for the state from the shore tourism revenue. The state's costs were about $18 million per year for the monitoring and improvement efforts. Therefore, an $18 million per year shore monitoring and improvement effort was protecting $900 million revenue for the state, providing an in-year payback of about 50 times our state's annual investment. Business would consider this a star investment and would be sure not to underinvest in a way that would risk killing the goose that was laying golden eggs. Of course, shore tourism revenue would not drop to nothing immediately if one year of shore water quality monitoring and improvement did not happen. It would take a few years of neglect to cause that.

Another way to look at the state's investment is to consider the incremental benefit from these programs. Before we started the shore-water quality improvement programs, New Jersey's shore revenue was declining or flat. Since 1990, the shore tourism revenue had been increasing at the rate of about $1 billion per year. That increase in revenue could be directly attributed to the improved shore-water quality. The state received about 6 percent of this increased revenue as tax income, which was an increase in state tax income of about $60 million per year. This incremental tax income is more than three times the annual investment of $18 million per year for the program that produced it, or about a $42 million net incremental benefit to the budget each year.

Therefore, the shore monitoring and improvement programs could be viewed as a great investment for New Jersey and they actually helped to achieve the objective of balancing the budget. In addition, these programs met another priority of the New Jersey voters who wanted excellent shore water quality they could enjoy.

The Budget Challenge

However, we had not yet improved the government funding process to consider these facts in their funding decisions. If we were to avoid continued lack of financial support for such a beneficial program, we needed to apply a quality improvement effort on the state government's budget process. The budget process should support a plan to reduce both our taxes and our shore water pollution. One of the challenges of management, in business or in government, is choosing between important priorities, when *both* need to be met. We needed a management approach that could meet both priorities. The required answer for many of the problems we face, such as the need for lower taxes *and* a better environment, is we need *both*, not simply a choice between the two. This also goes for education, health care, and many more of the pressing issues in America. We need a management approach that uses quality principles to address these issues to achieve both higher quality and lower costs.

While it may be harder to implement a quality approach than to simply make trade-off decisions of cost versus quality, the benefit of getting both lower taxes with a balanced budget while improving the quality of life in the community makes the extra effort worthwhile.

While the shore monitoring continued in 1994, the 1994 budget cut eliminated the state support for our core improvement program, which addressed the major root cause of shore pollution. The cut in the Sewage Infrastructure Improvement Act was a setback of not only our financial support for the four shore counties and 90 shore municipalities, but also was another emotional setback for our volunteer team. Our team took another look at Ed Norton's saying: "Once you put your boots on—you've got to go all the way." We began to realize that we needed our boots on as much in the budget process as we did in the sewage process.

LEADERSHIP FOR EXCELLENCE

Even though the state funding was a relatively small part of the total effort, we feared the budget cut would cause many municipalities to cancel their planned improvements due to insufficient funding, or cancel because of a lack of state-level commitment to improving the shore. If the state wasn't committed and other shore towns also reduced their commitment, why should any shore town bother to continue? It would be ineffective for an individual town to continue improvement efforts, if other towns and the state were not. This required a team effort by all the shore counties and municipalities—and the team required leadership.

The leadership role in improvements that require a critical mass of participants, such as all the shore towns, is critical and appears to be inadequately understood by the state and federal governments. Our team now realized the difficulty that this lack of leadership creates, in New Jersey, in other states, and in the federal government. The state needs to provide leadership to encourage the required critical mass of towns to achieve excellence and the federal government must provide leadership for the required critical mass of states to achieve excellence.

It is not the highest level that creates excellence. It is the next level down that creates excellence for any level of business or government. Excellent towns create an excellent state. Excellent states create an excellent country. Excellent business units create an excellent company. Therefore, the primary job of the higher level is to provide leadership and support to the next level down. Excellent countries lead and support states. Excellent states lead and support municipalities. Excellent companies lead and support business units. Excellent school districts lead and support excellent schools.

The 1992 budget cut had eliminated the state shore monitoring program that summer. This second budget cut in 1994 eliminated state support for the key shore improvement effort. Together these two major state budget cuts in the support of our shore monitoring and improvement program put the credibility of the state leadership role for shore improvement in question. Our QNJ environment team and the shore counties and shore municipalities could not count on the state for support. The state had now let them, and our team, down two out of three consecutive years.

Our QNJ environment team had a double problem. We were below "F" performance without even minimum goals and our beachgoers wanted "A" level performance. The citizens and the team had goals, the state did not.

Our difficulty now seemed to be with the support processes (for example, government funding of the core process work). One of the powerful parts of the Baldrige National Quality Award criteria is its assessment of both core processes and support processes of an applicant. In fact, I believed the AT&T support divisions' willingness to apply the Baldrige criteria assessment to their processes (human resources, real estate, information management systems, finance, etc.) was a key to why two AT&T units had won Baldrige awards in 1992. You have to address both core and support processes to reach the levels of excellence required to win the Baldrige Award. Our QNJ environment team had not yet reached Baldrige level because, while our core processes were approaching good to excellent, our support processes were still subpar. An example of the lack of support process excellence was the two budget cuts in our program.

HOW IS THE QUALITY MANAGEMENT APPROACH DIFFERENT?

A quality management approach differs from the government's usual law-focused management approach to solving problems in several ways:

1. Stretch goals are set based on customer/voter expectations vs. no goals or minimal performance goals are set.
2. Root causes of problems are identified and prioritized using data, and successful countermeasures are shared and replicated vs. special interests and emotional or political problems define prioritization.
3. Solutions are developed by many including those who have to implement them vs. a few smart people in Washington develop the solutions.
4. Funding of solutions is based on quantified cost/benefit analysis vs. funding of solutions is based on emotional or special interest influences.
5. Recognition is given for new local improvement programs vs. punishment for not following old regulations.
6. Through recognition, best practices are shared with others vs. a focus on the worst practices.

Punishment or Recognition?

Clearly, punishment is required for not meeting minimal performance levels acceptable to society. However, in the case of shore pollution problems, the causes lay primarily with the inadequate sewage and storm-drain systems, which are the government's responsibility, and government does not punish itself. Closing a beach to the public when sewage is not treated properly by a municipality does not punish the people responsible for the pollution.

The daily actions of a majority of U.S. citizens are not motivated by the fear of punishment. This includes those who wake up planning to do no wrong and many of those who wake up planning to do wrong. We can see this every day on the drive to work. Many drivers continually operate in the D– to F+ zone. We can also see it by simply reading the newspaper or watching the nightly news. The legal system is not a good motivational system for most people in the United States. It may prevent the majority from falling from a D– to an F, but it does not prevent a large number of people from operating in the F range.

I recognize the need for deterrents, for laws and regulations, to prevent chaos, to prevent the majority from falling from D– to F (Table 7-1). However, we also need a positive motivational management approach that encourages people in the United States to achieve excellence, to perform in the A and B range (Table 7-2). We need both. Again and again, we find the answer to many of our problems is *both*, instead of a simple choice between two alternatives.

We need both positive motivation and deterrents. We need both lower taxes and higher quality. We need both quality core processes and quality support processes. We need both minimal standards and stretch goals for excellence. We need government and business and citizens working together to address our community problems with a quality approach to achieve both higher quality and lower costs.

People are motivated by setting appropriate stretch goals, by data on the real root causes of problems, by being empowered to develop their own solutions (with the necessary support, sharing of what works and recognition from others) by being accountable for their own results, by striving to be the best as measured by the

Table 7-1.

Regulation-Driven Compliance with Negative Motivation		
Phase I **Regulations**	**Phase II** **Monitoring**	**Phase III** **Punishment**
• Identify problems • Create regulations • Solutions based on politics, influence and wisdom of staff	• Implement rules and regulations • Monitor compliance	• Punish those who violate rules and regulations • Fines, etc.

Table 7-2.

Goal-Driven Process Improvement with Positive Motivation		
Phase I **Vision/Goals**	**Phase II** **Empowerment**	**Phase III** **Recognition**
Create a vision and emotion to close gap • Goals • Measure	Support and mentor those who must improve • Benchmark progress • Education • Data analysis/data-driven solutions • Process improvement	Recognize and renew champions who improved toward goal Share • Awards • Rewards • Successes

highest levels of performance, and measuring ourselves against criteria for excellence.

Examples of awards for excellence are the Olympics, the Malcolm Baldrige National Quality Award in the United States, and our QNJ New Jersey Shore Quality Award at a state level. These three awards motivate achievement of excellence on a personal, business, and community level, respectively.

This same approach could be expanded to the critical areas of education and health care. Proposals for expanding the Baldrige Award to also include the education and health care sectors was initially discussed in 1994. However, the Baldrige Award Foundation, supported by business, and our federal government were both having some financial problems in 1994 and were not ready to take on the expansion of the Baldrige Award to new sectors. In 1995, the Baldrige Foundation, chaired by Bob Allen, reconsidered its capability to support the education and health care sectors and decided to initiate plans to expand support to those new award categories.[1]

Recognition Shares Success

The team was having success applying a quality approach to our New Jersey shore community problem. We had achieved a 90 percent reduction in beach pollution, shore tourism had the highest growth rate of any sector in the state, New Jersey had moved ahead of Hawaii in total shore tourism revenue, and the dolphins were back. However, we had a continuing problem obtaining government funding for our core processes; therefore, we needed to apply our quality approach not only to our core processes but also to our support processes (the legislation and budget processes). While we had begun to apply a quality approach to government support processes (by determining root causes of budget cuts to high-benefit programs) we had not yet implemented process improvement solutions (countermeasures) sufficient to prevent future recurrence.

When I gave my talk in Washington, people at the Baldrige Quest for Excellence Conference seemed pleased with our success in using a quality approach on a community problem. They also recognized that we had not yet solved the support process problems.

IMPROVING QUALITY WHILE CONSOLIDATING AT AT&T

Our largest AT&T units, consumer communications services and the network services division, were planning to apply again for the Baldrige National Quality Award in 1994. The leaders of both units had changed, and both new unit heads decided to continue the quality effort toward the goal of excellence defined by the Baldrige award. This continued support of quality improvement after a change in the top management was something of which I was very proud. This was an indication that our quality approach was becoming part of the way AT&T was managed, becoming part of our culture, not just an initiative led by an individual leader. Of course it had helped that the new head of the AT&T communications network division was Frank Ianna, who had been promoted from his position as chief quality officer for our communications services. Frank also kept his function as chief quality officer for our AT&T communications services in addition to his new leadership role for the AT&T communications network planning, development and management.

Both units had made a number of process improvements since the prior year. To address increasing competition, a new marketing program had just been introduced. Other changes to further improve our cost and performance for customers were underway, including office consolidations, and major investments in network technology.

The conversion from analog technology to digital technology was nearing completion in the industry. Throughout the telecommunications industry new technology was reducing some jobs, and competitive pressures required continuous cost reductions with continuous quality improvements. A number of special human resources programs were also being introduced to help people to deal with the transition associated with office consolidations and the introduction of new technology.

The Baldrige application team prepared their new application and submitted it in April 1994 confident that we had improved. But we had reservations about our ability to win an award in an industry that was consolidating.

QUALITY STANDARDS FOR A GLOBAL MARKETPLACE

On a worldwide basis it could be seen that, in the long run, the high-quality low-cost countries were winning in the marketplace, which enabled them to provide the highest levels of employment. Western Europe was at 12 percent unemployment, the United States at 6 percent unemployment, and Southeast Asia was at 2 percent to 3 percent unemployment in 1994.

The Southeast Asian countries—Hong Kong, Singapore, Taiwan, South Korea, Malaysia, Indonesia, and Thailand—followed Japan's lead by using a quality approach to management that relied on educated workers providing improvement ideas that were acted on by management to continuously improve the capability of their businesses. The United States was next in line with the electronics and automotive manufacturing industries working hard to catch up with Japan. Europe was far behind, trapped in its government-controlled system of management. For example, European governments were just beginning to plan for the sale of their government-owned monopoly telecommunications companies to the public. Competition in the telecommunications industry had not been allowed in Europe, because the government owned the telecommunications companies and the government made the rules.

Low Standards Mean Low Expectations

The Southeast Asian countries were providing an inviting climate for private investment and they were also investing more than the United States in primary and secondary education. More people using their brain was what the information age required to increase the competitiveness of a country. Competition was now on an international scale, and both countries and businesses were learning to win using a quality approach.

However, a decade after "A Nation at Risk," a grim report about education in the United States, little progress has been made. The United States has the shortest school year of any industrialized nation in the world and has no national means to measure what students are learning. As with the ocean-water quality problem, some at the federal level must feel that if we do not measure it, no one will notice how we are really doing. People who see the schools or the oceans firsthand know how we are doing, but a measure and a goal for improvement would help us do better.

Perhaps one of the best measures we have of what U.S. students are learning in school is the SAT college-entrance exams that are taken on a voluntary basis by many high school students and administered by an independent organization. In 1994 it was announced that the average SAT scores in the United States had been declining for the past four decades. In 1941, the average score was 500 in both math and verbal skills. In 1994, it was 478 in math skills and 424 in verbal skills. To address this decline, the College Board officials announced they would "recenter" the scale, which meant that by answering the same number of questions correctly a student would get about 20 extra points on the math exam and about 80 extra points on the verbal exam, beginning in 1995. Recentering scores seemed to be a simplistic approach to fixing our secondary education results. Simply change the measure to add points to the results. If some time in the future we find runners slowing down at the Olympics we could simply subtract seconds from their time to also make their results better. It certainly is easier than training harder, if the other countries would only do the same.

This "adjustment" to the SAT scores was justified, by the College Board officials, on the basis that since a higher percentage of secondary school students now take the test, we should expect results to be lower. However, since the SAT is taken voluntarily and kids don't like to spend a Saturday morning taking a four-hour exam, the United States must now have a higher percentage of kids who actually want to go to college. We also know that, in the 53 years since 1941, the Industrial Age of work in a factory has shifted to the Information Age, which requires a college education to meet today's job requirements. We also have 53 more years of advancements in the world to know about. These are hardly reasons to accept having a lower average score from those who want and need to go to college, or to cover up the lower scores by "adjusting" the measures.

Instead of changing the basis for the measure, the root causes of its decline should have been addressed to improve the education of those who want to and need to go to college. The measure acknowledged the fact that the average education of those who go to college is declining in the United States. A declining education should not be acceptable. Once you cover up the problem by adjusting the measures, the power of having a gap to close is lost, reducing the motivation of many. If you don't have a measurable gap to close on a serious problem, why improve? By adjusting the SAT scores, we have lowered the standards expected from the U.S. secondary education system, when perhaps we needed to raise them.

Japanese Standards for AT&T Management?

One of our AT&T manufacturing units was concerned that perhaps the Malcolm Baldrige Quality Award criteria were set at a lower level than the Japanese criteria. They were worried that even if they won the Baldrige Award they might not be able to beat their Japanese competitors in the global marketplace. Therefore, they set their sights on winning the Deming Prize, Japan's national quality award. No U.S. manufacturing location had ever won Japan's quality award, even though it happens to be named after a leading U.S. quality professional, W. Edwards Deming.

Our AT&T U.S. manufacturing applicant for the Japanese Deming Prize was expecting to be ready for their examination later in 1994. The application was reviewed with our AT&T unit quality directors early in the year, sharing what we learned from using the award quality criteria in one of our better units. One of the most important things we learned was the higher expectation set for the quality knowledge of senior management. The Japanese did have a higher standard than the U.S. criteria for senior management's knowledge on how to manage with a quality approach. The entire senior management team had to completely understand their quality system for managing including the key processes and the key quality tools used in their quality approach. The examination of the senior management team by the Japanese examiners was much tougher.

Both Baldrige and Deming examiners would be coming to AT&T units later in 1994 to check on the quality systems at our leading units.

ALLOCATION OF RESOURCES AT THE SHORE

The summer of 1994 started with good weather and good water. Based on the great year we had in 1993, the shore rentals were sold out early in the year and people were predicting continued increases in shore tourism.

In spite of budget cuts in other areas, in 1994 New Jersey decided to fund one project 36 years after the U.S. Congress had authorized it in 1958. The project was to build a shore buffer to protect a portion of the Jersey shore from coastal storms. The portion of the shore to be restored was a 21-mile stretch of beach for a dozen towns along the northern part of the Jersey shore. The Army Corps of Engineers would pump millions of cubic yards of sand onto the beaches from Sea Bright to Ocean Township. The buffer built by the new sand would result in wider beaches that would slow down erosion of the coastline and prevent or delay homes from being washed into the ocean.

Since Mother Nature does not let sand stay where you put it, the project would require periodic replenishment. The total estimated cost over 50 years was $1.84 billion to regularly pump sand onto this portion of the Jersey shore. The federal government agreed to pick up the tab for $1.19 billion and New Jersey would have to pay "only" the remaining $651 million, of which the state would pay 75 percent and the shore communities 25 percent.

The cost to the state in 1994 was only $15 million, funded by a Shore Protection Fund proposed by Governor Jim Florio and passed by the state legislature in 1993. Finding the $15 million for pumping sand on a few beaches was no problem for the same state that could not afford $6 million to fund the Sewage Infrastructure Improvement Act that supported keeping all 127 miles of our ocean and bay water clean from pollution. I couldn't understand why the state government, 36 years after a federal bill was passed, came to the conclusion that it could spend hundreds of millions of dollars over many years putting sand on a small portion of Jersey's beaches, but lacked the discipline required to continue improvements in preventing pollution of our entire shoreline.

I also couldn't understand why many shore environmental groups were more focused on stopping the dredging of the New York–New Jersey harbor than dealing with the sources of shore pollution.

I tried to ignore these sand replenishment and dredging efforts, in spite of our lack of funding support in 1994, and focus on what we could do to prevent the root causes of New Jersey shore water pollution. However, these two major shore projects were hard to ignore because they got regular media coverage. In June 1994, the beach replenishment project had to be temporarily halted at the Monmouth shore when the Army Corps of Engineers uncovered 41 artillery shells ranging from 3 inches to 10 inches long, including one live round. Apparently the sand pit they decided to pump sand from was also the test site for rounds fired from the former army installation on Sandy Hook. Ammunition was tested on the north beach from 1870 to 1919. Ammunition shells were probably the one type of litter that we hadn't previously found on New Jersey's shores, until the Army Corps of Engineers decided to help us out.

While the reports of water quality remained good, as did the tourism, the actual beach-block-days of closures due to pollution in 1994 did worsen. The 88 beach-block-day closings in 1993 increased to 211 in 1994, the worst results we had since 1990. It seemed clear that the withdrawal of funding support for the Sewage Infrastructure Improvement Act had the negative impact on the community improvement efforts that we had feared. Once again, I was not looking forward to the QNJ annual quality conference, where I reported the annual shore water quality results.

AN AT&T TRIPLE

While September closed the summer months with less than desired results for our Jersey shore improvement efforts, October brought some good news on the AT&T quality improvement efforts. All three AT&T units applying for national quality awards received the good news that they had met the standards of excellence required.

The AT&T consumer communications services unit and the network services division combined to win the Malcolm Baldrige National Quality Award. The AT&T microelectronics power systems unit won Japan's national quality award, the Deming Prize. In addition, an Orlando, Florida AT&T microelectronics factory had also won the U.S. Shingo Prize for Manufacturing Excellence. (In 1992 the AT&T microelectronics power systems unit had won the U.S. Shingo Prize.)

This triple quality award recognition was cause for celebration at AT&T. The two AT&T units that had won a Baldrige award in 1992 represented less than 15 percent of AT&T's revenue. But AT&T was the first company to win a double Baldrige award (two units from the same company). For all the Baldrige awards ever given, from the initial awards in 1987 through the most current 1997 awards, AT&T, Solectron Corporation, and Xerox are the only companies to have more than one unit win. The 1994 consumer communications service unit winner represented close to 40 percent of AT&T's revenue and the network services division was the core unit in AT&T that provided the worldwide network support for all of AT&T.

President Clinton was unable to attend the 1994 Baldrige Award ceremony in Washington, D.C. due to foreign policy meeting conflicts. Vice President Gore presented the awards to the winners and said some very nice things about the ability of a large company like AT&T to apply the quality principles that provide broad support for employee education and empowerment. I felt very proud of our winning AT&T units and that the many years of hard work by thousands of our people had been recognized. These recognition moments are very important. They recognize completion of one phase of work and the beginning of the next phase of continuous improvement.

We had plenty to celebrate at our annual AT&T quality conference, and we celebrated by recognizing these winners and the improvement made in many of the other AT&T units as measured by our internal Chairman's Quality Award, which was run using the Baldrige Award criteria.

Business Excellence

In 1994, the chairman introduced a new AT&T award called the Business Excellence Award (BEA). This award was based on the results a unit achieved compared to industry performance, on key measures for customers, owners and people. This new award set a new goal to meet a new level of excellence for all three of our key stakeholders. This new and higher standard of excellence was required to motivate continuous improvement at AT&T after having won several national quality awards (Figure 7-1).

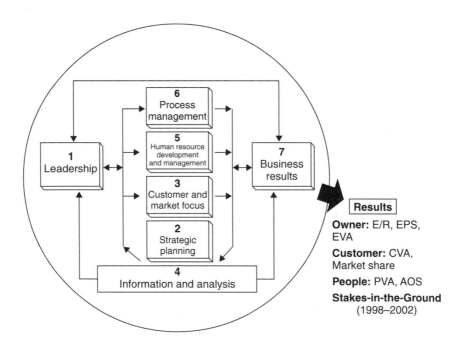

Figure 7-1. MBA/CQA Management System Model

The measure for customers was customer value added (CVA). This measure is a measure of how well AT&T satisfies customers compared to how well our competitors satisfy their customers. The measure is obtained by surveys of both our customers and our competitors' customers conducted by independent market research organizations. While the surveys ask a number of questions, a key question is: How would you rate the service you received on a "worth what you paid for it" basis? This question gets at the overall service value, including both quality and price factors. We have found that the response to customer satisfaction questions, when compared to the response received by customers of our competitors, with proper weighting, can be correlated to subsequent purchasing behavior. However, for units with large market share, like most of our AT&T units, the requirement is to provide a significant value differential. Being as good as, or even just a bit better, isn't good enough for AT&T units, we have to be significantly better than our smaller competitors. Good isn't good enough! AT&T must be excellent.

The key measure for our investors was economic value added (EVA). This measure provides an indication of the amount earned on our invested capital above the cost of money. EVA is the amount earned above the cost of capital. A positive EVA can be used to expand or improve the business capability.

The key measure for people was people value added (PVA). This measure provides an indication of the work environment AT&T provides for our people using a 30-question annual survey of all AT&T employees.

The AT&T Quality Office compared the unit results for all AT&T business units on all three measures (CVA, EVA, and PVA) with the average and best-in-class performance in their respective industry.

AT&T was in four different industries in 1994: telecommunications services, telecommunications manufacturing (now Lucent Technologies), computer manufacturing (now NCR), and financial services. In 1996 AT&T completed a split of the company (a trivestiture), creating three new separate companies: AT&T for telecommunications services, Lucent Technologies for telecommunications manufacturing and NCR for computers. AT&T Capital

Corporation, which provides financial services to business customers, was also split off.

To be eligible for a Business Excellence Award (BEA) a unit had to be better than the industry average (top 50 percent of the industry) on all three measures and excellent on at least one of the three measures. Excellent was defined as being in the top 25 percent of the industry on a measure.

The BEA business results criteria included an above-industry-average balanced performance for all three key stakeholders and excellent performance for at least one stakeholder. Above-industry-average performance on two measures and excellent on one met the bronze BEA criteria. Above-industry-average on one and excellent on two met the silver BEA criteria. Excellent results on all three met the gold BEA criteria. These criteria proved to be very difficult. In 1994, the only AT&T unit to meet the demanding BEA criteria was our AT&T Universal Card Services unit, which had won a Baldrige award in 1992 and had continued their improvement efforts in 1993 and 1994.

Although our 1994 quality-award-winning units also performed at the excellent level on some of these key stakeholder measures, they fell just below the industry average on one, PVA. In our consolidating mode, employee satisfaction had fallen below average, which disqualified our leading quality unit from winning a Business Excellence Award. They had won a Baldrige award but did not yet qualify to win AT&T's more demanding Business Excellence Award. However, these units were close enough to winning in 1994 and were my bet to win the Business Excellence Award in a future year with continuous improvement efforts. In fact, our 1994 Baldrige Award winner did win the AT&T BEA award in 1995.

A comparison of each AT&T unit's Chairman's Quality Award (CQA) scores, based on Baldrige criteria, showed a direct correlation to the desired business (CVA, EVA and PVA) results. Our quality efforts, our investments in time and process improvements, were producing a "return on quality" for all three stakeholders. When we improved our management system, as measured by the Baldrige criteria (CQA), we were improving our capability to meet customer expectations (CVA), owner expectations (EVA), and people expectations (PVA), better than our competitors.

The 1994 results for our key stakeholders, relative to industry performance, was a confirmation of our belief that improved quality would produce improved business results.

Getting a Return on Quality

In 1994, AT&T had its best year in the decade since the divestiture of the Bell System in 1984. As a result, I received invitations to speak on "how to get a return on quality" at several conferences, including the World Quality Conference in Finland, the American Society for Quality Control (ASQC) conference in the U.S., and Conference Board conferences in Brussels and New York. With three Baldrige awards, two Shingo awards, a Deming Prize, several state quality awards, and great business results, the AT&T quality office also became an important stop for those benchmarking successful quality approaches, including many of AT&T's business customers. It seemed that AT&T had finally recovered from the divestiture 10 years before, and our success was being recognized. However, in the following year, we were to begin feeling the effects of major new legal and regulation changes in the telecommunications industry.

1994 QNJ SHORE QUALITY AWARD WINNERS

I spoke at the annual QNJ conference and explained how, for the second time, problems with a support process had led to an increase in beach-block-day closings on the Jersey shore. While I was again worried that the conferees might lose faith in the quality process as a result of this second setback, I was pleasantly surprised again by the supportive response from the conferees. They understood that any quality journey is not without setbacks—and often more than one.

In addition, I was most pleasantly surprised by the remarks of one of our role-model mayors from the town of Avalon as he accepted his 1994 QNJ Shore Quality Award. Mayor Martin Pagliughi said, "The shore municipalities and counties could continue their shore improvement work without continued financial reliance on the state." He shared information on both his core infrastructure improvement efforts and his financial support processes and sources of resources. Our role-model town for infra-structure improvement was also a role-model town for a financial support process. In 1995, our QNJ group needed to share both his core process and his support process with the other shore towns.

Lessons Learned (1994)

1. A cost/benefit analysis should be used to prioritize major improvements.
2. Withdrawal of financial support for best practice use results in less use of best practices.
3. Government doesn't regulate itself very well.
4. World competition requires world class standards of excellence.
5. Adjusting measures to lower standards does not motivate achievement of excellence.
6. Quality excellence produces business excellence for all stakeholders.
7. The quality journey is not without setbacks.
8. Excellence must be led and supported.
9. More often than not we require both, not a choice between two requirements.
10. The Japanese have a higher standard for senior management knowledge of their quality approach.

CHAPTER EIGHT

A Factor of 10 Improvement
(1995)

A WORLD ECONOMY KEEPS PRESSURE ON IMPROVEMENT

Early January government reports shared the good news that 1994 had been a year of great job creation for the United States. In 1994, more jobs had been created than any other year since 1984. There was still a great job transition happening as we continued a shift from a U.S.-based agricultural and industrial age to a global agricultural, industrial, and information age, but the net result in 1994 was positive.

In 1994, the U.S. economy produced an increase of 3.5 million jobs, which reduced the U.S. unemployed by 1.5 million, with the remaining 2 million jobs going to new entrants to the job market. U.S. unemployment dropped from 6.7 percent in December 1993 to 5.4 percent in December 1994. Through June 1997, in spite of increased global competition and NAFTA trade agreements, the United States had not only sustained this improvement but reduced unemployment to under 5 percent.

Another impressive result in 1994 was U.S. competitiveness in the manufacturing industry. In 1993, the United States lost 130,000 jobs in manufacturing, but in 1994 the United States added 300,000 jobs in manufacturing. Our efforts to compete on a global basis, using a quality approach, had begun paying off. In addition, in spite of the low cost of labor in other nations, U.S. wages were still able to increase by 3 percent in 1994. Through 1996, the United States continued annual salary increases at a 4 percent level, with a projection for 1997 to remain at the 4 percent level, enough to sustain current standards of living during a period of low inflation and tough world competition.[1]

Unemployment in New Jersey in 1994 improved in line with the national improvement. However, New Jersey at 6.1 percent was still above the US average of 5.4 percent, while lower than California who finished the year at 7.4 percent. The end-of-year 1996 results showed that New Jersey had also maintained the 1994 improvement in unemployment, finishing the 1996 year at a 6.2 percent unemployment rate, and reaching a 5.5 percent level by the end of March 1997.

The U.S. economy in 1994 was showing indications of an improved capability to compete on a global basis, to hold jobs and current standards of living. However, the United States still had a major trade deficit and a major federal government budget deficit that was supporting current standards of living at the cost of future standards of living. The United States had improved, but not yet sufficiently to retain jobs, hold standards of living, and wipe out trade and budget deficits. We still needed further improvement to avoid a lower standard of living now, or in the future.

We began to see the capability of open, global competition to prevent any large price or salary increases, thus reducing the probability of business cycles causing large changes in the rate of inflation or interest rates. In 1995 and 1996, it seemed that every morning on CNN's business news I would hear a financial analyst trying to explain our multi-year run of low interest rates, low inflation and modest growth, giving much of the credit to actions by Alan Greenspan at the Federal Reserve. However, this is the same Federal Reserve we had in place during the periods of large business cycle swings. The Federal Reserve reacts to the economic changes in the country, more than it creates them. The increased global competition and the improved capability of the United States to compete in a global economy are the underlying reasons for the reduction in our economic swings. Of course, the Federal Reserve still has the power to raise interest rates whenever they feel like tinkering with the economy, which causes higher rates for loans and mortgages, thereby contributing to a cost of living increase for many who have the need to borrow or who have adjustable mortgage rates. Interest rate increases also reduce purchases, reduce investments, reduce jobs, reduce national productivity, increase trade deficits, and increase federal government deficits due to reduced tax revenue from fewer jobs combined with increased government payments to the unemployed.

The Federal Reserve claims that it raises interest rates to protect U.S. citizens from potential increases in wages and prices that can result from low unemployment. This thinking was developed in a period when the U.S. economy was isolated from the effects of world trade and world sources of labor. However, in today's world of increased open global competition, it is unlikely that we will again see the large annual price and salary increases that cause high inflation. U.S. businesses must compete in a world competitive situation where business in countries like Japan can have access to investment money with a prime rate of 1 percent or 2 percent. The more a country opens its doors to global competition the more it lets in a natural control for high inflation and high interest rates. In a world of global finance, perhaps some day the Federal Reserve will stop tinkering with the U.S. economy based on old U.S.-based economic models and understand the global economic model that requires the United States to be as competitive as possible in the world markets with the lowest interest rates and the fullest employment possible. It is with a full work force working that the United States has the greatest chance of sustaining its standard of living today and in the future. The global competition in products, services and labor will be the natural control on wage and price inflation; we no longer need the Federal Reserve to do that job.

Competing Globally

On the other hand, the more global competition we allow in our country the more we must be able to compete on a price and quality basis to maintain our jobs and our standard of living, and establish a balance of trade and a balanced budget.

Many other countries, such as China, are starting out with much lower labor and tax rates than the United States. However, they aim to improve their own standard of living. As a country succeeds in competing for a slice of the global economy, as China has been doing and will continue to do, it tends to raise its standard of living and reduce the labor and tax advantage they have over the United States. This is what happened in Japan over the last 30 years.

Global competition results in business relationships that reduce the likelihood of military conflicts. Many nations have significant economic differences. Trading partners, who are allowed to compete for economic improvement, are less likely to go to war to achieve that same economic improvement.

Global competition is good as a natural control for inflation and interest rates and it provides a peaceful means to enable the standard of living to be improved throughout the world, including countries such as China. However, it is a threat to U.S. jobs and our standard of living if we can't compete.

Global competition is going to continue to increase in the United States and in other countries of the world at the same time. We are in a global economic war with every other country in learning how to compete for business here in the United States as well as in other countries, to maintain our jobs and our standard of living. The United States hasn't always been better off than the rest of the world and there are no world rules that say that the U.S. has been granted the right to a higher standard of living than others.

The United States must do better than the best in other countries, where many are operating on a lower standard of living, to

Bruce Springsteen, a Jersey boy, has sung for the past 30 years about the misfortunes of many in the United States who have lost their jobs, or their standard of living, even though they worked as hard as their parents had. It doesn't seem fair, and it may not be fair. However, working as hard as our parents did is not good enough to maintain our jobs and standard of living. People in Michigan in the auto industry have recognized this hard cruel reality, and people in all sectors of the economy must also recognize it. Springsteen's songs recognize the plight of those caught in a declining or losing industry. They produce a sympathy we all feel for those in his songs. We must learn how to compete on a global basis to avoid ending up being a character in a future Springsteen song.

win in the world trade battle. History is not the place to get benchmarks for performance in business, education, health care, government, or any sector of the economy. The United States must benchmark itself against the best in the world today. The U.S. must compete for jobs and for our standard of living in a global economy, where the best will win.

HISTORY OF THE QUALITY APPROACH

After World War II, while the rest of the world had to rebuild, the United States was able to develop the world's most powerful industrial economy and achieve a great standard of living. During the years of my youth, 1945 to 1965, it seemed that the United States was able to do anything it wanted to do: build a strong industrial economy with a growing number of jobs, raise the standard of living, increase national defense spending and go to the moon all while balancing the federal budget and achieving a balance in trade. When I graduated from college in 1966, as an electrical engineer with average grades, I had five great job offers to choose from. Today's engineering graduates, including a nephew of mine, may require six months of job searching to get one decent job offer.

During the post-World War II years, other countries like Japan rebuilt their industrial capability and they used a quality approach to produce higher quality products at a lower cost. The United States was not focused on the new global competition that it would begin to face in the late 1960s. The United States underestimated the competitive capabilities of other countries and didn't bother to learn this new approach that could produce higher quality at a lower cost.

Japan began taking over the industrial leadership position. They started with motorbikes and proceeded to dominate entire industry sectors such as consumer electronics. From cars to VCRs, Japan's product quality and cost competitiveness was dominant. The United States lost the entire consumer electronics industry to Japan. While U.S. citizens were watching their TV sets, the Japanese were making them. The only way the United States saved its auto industry was to legislate a quota on Japanese imports, which gave the U.S. auto industry the time it needed to improve its quality and cost of operations, and the time needed to learn and deploy a quality approach. Quality is now "Job One" at Ford and at other U.S. auto manufacturers.

Many of the countries in Asia have learned the quality approach that worked for Japan, and they are also now strong in the global competitive marketplace. While the quality approach was expanding in Asia, and being used by leading companies in the United States, the business media in the United States continued to run articles that treated a quality approach as a fad rather than as a necessary part of a successful business management system.

While the United States was focused on fighting, and winning, the Cold War with Russia, it was also fighting, and losing, the economic war with Asia.

A U.S. QUALITY APPROACH TO IMPROVEMENT IS ENCOURAGED

In 1985, the American Society for Quality Control (ASQC) initiated an annual Quality Month campaign, to be held in October each year, to increase awareness of the need for the United States to increase use of a quality approach to win the economic war.

AT&T was the first sponsor of this ASQC Quality Month campaign, which caused AT&T to renew its quality policy. For the past several years, AT&T has sponsored the ASQC National Quality Forum October satellite broadcast, which is shared nationally with ASQC sections, and business, government, and education sites that have a capability to receive the broadcast. (In 1997 the ASQC changed its name to ASQ.)

To further encourage the use and sharing of best quality practices, in 1987, the federal government established a national quality award named after Malcolm Baldrige. Baldrige was a popular and outgoing Secretary of Commerce, who was killed when he fell off his horse while riding. Business agreed to join in support of this national quality program by allowing volunteers to be examiners, by establishing a Malcolm Baldrige Award Foundation, and by having winning companies openly share their best quality practices with others. This created a government and business partnership for recognizing and sharing best quality practices. This partnership is one that Japan had established about 40 years earlier with its business community and the Deming Prize, Japan's national quality award.

Three AT&T business units, and 29 other U.S. companies, won a Malcolm Baldrige National Quality Award between 1987 and 1997. In 10 years, 32 companies, or major business units of large companies, had met the demanding criteria and examination process required to win a U.S. National Quality Award. More importantly, these companies, and many other companies, had learned and deployed a quality approach required to compete successfully in the 1990s. An investment in the stock of a Baldrige-winning company has consistently produced a return two to three times higher than the S&P 500 index companies, which is one indication of their management system's capability to deal with the many factors that cause a particular company to do well or poorly at any given point.[2] Of course, at any time, a particular company in either the Baldrige portfolio, or the S&P portfolio, may have a below average year.

EDUCATIONAL SUPPORT FOR A QUALITY APPROACH

Japan used a quality approach to improve their processes and engage their employees in a journey of continuous improvement. U.S. quality gurus brought the initial quality approach to Japan at the request of the U.S. government, to help improve Japanese telecommunications services after the war. Japanese quality leaders took those initial ideas and built upon them. Japan's business leaders learned and used this quality approach to rebuild their industrial power. The United States had been winning without a quality approach between 1945 to 1965; therefore there was a resistance to teaching or learning this new way of management. In Japan, to survive as a country, they knew they needed a way to be better than the United States in manufacturing, given their limited natural resources and their loss of World War II.

In 1993, after AT&T had won two Baldrige Awards, I worked with our AT&T college recruiting department to do a survey of U.S. colleges' quality curriculum and their interest in learning more from industry quality leaders. Even in 1993, there was still very little being taught in our universities about a quality approach. However, there was an interest in about one-third of the universities contacted to begin learning about the quality approach used by leading U.S. businesses. In response to this university interest, I worked with the AT&T human resources department to create an AT&T Driving Quality on Campus program. This program provided an experienced AT&T speaker to give a talk on the AT&T quality approach that had helped us win two national quality awards, to the administration, faculty, or students of any university when requested. During the three-year period from 1993 through 1995, AT&T managers gave hundreds of talks at universities on our AT&T quality approach.

I had the opportunity to speak about our AT&T quality approach at a number of universities including Pennsylvania State University, MIT, Harvard, Boston University, George Washington University, University of Maryland, Polytechnic University in Brooklyn, Dominican College, and the school I graduated from, Northeastern University. Each of these opportunities gave me a chance to share our AT&T quality approach and explain what quality knowledge future business leaders would need as part of their education, in order to compete in today's global economy.

As AT&T's quality vice president, I was, in effect, the quality dean at AT&T. The job included providing the required quality knowledge in the form of books, courses, and internal consulting support—knowledge about quality that the education system had failed to provide. In fact, the quality-support job in business often requires educational rework because the education system failed to include this important knowledge in the curriculum. In quality terms, a portion of the quality-support role is a cost of quality, caused by not providing a sufficient education in the first place. In addition, while the U.S. higher education system was the best in the world, U.S. businesses, led by people educated in those universities, were losing in the global economic war. The message was not only a sharing of our best practices, but also a plea for help. Business needed a total quality approach included in the curriculum of higher education institutions to reduce the need for quality education efforts on the job.

Universities needed to take some responsibility for the fact that the people they were educating were losing. The United States has a good education system, but it is not sufficient. The education system must include required quality courses in the curriculum of every academic major, along with required courses in math, literature, history, and computers.

While giving one of these quality talks to the administration department heads at MIT, the president of the university joined us for the last half of the session. I could never have gotten into MIT as a student, and it was unbelievable to me that I was at a blackboard explaining something to the president of MIT and his staff. I believe that the higher education system in the United States is the best in the world, a belief confirmed by the people I meet on my university visits. However, it does have a missing piece, a quality piece.

Corporate Quality Partnerships

In 1994 I was asked to share the work of our QNJ environment focus group with a New Jersey higher education commission chartered with improving the quality of higher education in New Jersey. As a result of that presentation and discussions on how to apply a quality approach to higher education institutions in New Jersey, the commission recommended that the 42 institutions of higher education in New Jersey each establish a corporate partnership to learn the quality approach.

AT&T and Johnson & Johnson (J&J) both agreed to establish quality partnerships with Rutgers. Rutgers is the largest school in New Jersey, with five campuses and almost 50,000 students. Rutgers is also the third largest university in the nation.

The AT&T–Rutgers quality partnership is one of about 30 major business–university partnerships in the nation. The partners in this program share their results at an annual National Quality Forum. They also share the lessons learned from their quality partnership efforts to improve the quality of higher education. In addition, the corporate sponsors of these partnerships have provided funding for grants, administered by the National Science Foundation (NSF), to university professors for study of corporate quality programs. These studies can be the basis of developing a quality curriculum to learn and share with others the best approaches to continuous business improvement. It became clear that the leading universities are research universities that research a subject before they create a course in it. Therefore, the research funding is a prerequisite to quality education in the classroom.

SAME RULES ON HOME COURT OR AWAY GAMES

To compete in other countries (away games) the United States will need the same free, open market that exists in the United States (home games). Today, no country has as open a market for competition as the United States. I am not talking about closing our markets, I am talking about opening up other countries' markets to the same degree that ours is open.

The U.S. government must negotiate with others to achieve a level playing field for global competition, with the same rules for home and away games. The U.S. government must begin to consider what is happening in the world, not just in the United States,

---

An Unfair Playing Field

In the late 1990s, some countries began to open up their long-distance communications service to open telecommunications service competition. In these countries, such as England, Canada, and Mexico, AT&T began providing service to one customer at a time to begin building a competitive long-distance network. However, in each country outside the United States, AT&T must compete with a national company that has a monopoly on local service and control of the local access to customers. In the United States, AT&T was split up in 1984 to separate the local and long-distance services to allow competition for long-distance by companies that did not have this local service monopoly advantage. However, the United States was the only country to cut up its national telecommunications company in this way. The open market enables a major foreign competitor to simply come in and buy telecommunications companies to establish a major market presence in the United States, while maintaining its monopoly of local service in their own country.

While our government authorized foreign monopoly telecommunications companies to merge resources with U.S. long distance companies the FCC simultaneously opposed the merger of U.S. telecommunications company resources to enable competition with these world players, thereby creating an unfair playing field both at home and away in the telecommunications industry.

when making rules that affect U.S. industries' ability to compete. Other countries are focused on creating an advantage for their businesses, and for jobs for their people, by creating an advantage for their businesses on their home court. They are often quite surprised when the U.S. government also creates an advantage for them in the United States as a result of a focus on regulating U.S. industry without applying the same rules to foreign companies.

U.S. businesses will require partnerships with businesses in other countries to sell and service products as well as to partner in investments required to bring U.S. products and services to market in other countries in order to serve global customer needs.

IMPORTANCE OF THE INFRASTRUCTURE TO PERFORMANCE

The concept of recognizing the importance of a country's infrastructure (that is, telecommunications, transportation, education) is now beginning to be recognized as a key to growth in the developing countries. It should also be recognized in the developed countries as a key to sustaining our jobs and our level of living in an increasingly competitive world. AT&T's performance as an efficient and effective telecommunications support for the nation actually helps the nation improve its own productivity and ability to compete for jobs in a global marketplace.

AT&T's U.S. business customers had a similar requirement to our New Jersey beachgoers: they both wanted zero outages or closings. If AT&T had a network outage, it should be restored in seconds, not hours—and if New Jersey had a beach closing, it should be reopened in hours, not days. Delivering this level of performance helps U.S. businesses be more competitive and New Jersey businesses to increase their tourism revenue—and open, clean beaches enable New Jersey residents to have more fun.

Other areas not yet adequately recognized as key parts of a country's infrastructure, both in developing and developed countries, are the systems and processes necessary to protect the environment, while sustaining growth. In the short run, a country can, and typically does, abuse its natural resources to create growth at a lower short-term cost. To sustain growth, however, a developed nation also needs infrastructure systems and processes that provide environmental protection. These systems and processes are those that prevent the land, water, and air pollution problems that can result in a degradation of the quality of life in a country, and therefore undermine the ability to sustain growth.

While Japan focused on developing its industrial power in the 1960s, 1970s, and 1980s, it failed to invest in a balanced way in its infrastructure. For example, in the 1990s, half of Japan's population lives in homes that are not connected to sewage systems. Therefore, in the late 1990s Japan is trying to compensate by spending 6.6 percent of its gross domestic product on public works projects compared to 1.6 percent in the U.S. (The 1.6 percent number in the United States is probably not enough to maintain its infrastructure adequately.) This late Japanese effort to catch up on its public works projects has resulted in a deep deficit in their own federal budget, even though they continue to have a very positive balance of trade. In 1996, Japan's central and local budget deficits amounted to 7 percent of gross domestic product, while the U.S. federal budget deficit is less than 2 percent.[3]

SUSTAINABLE GROWTH IN NEW JERSEY— DEALING WITH ENVIRONMENTAL ISSUES

Continued growth without environmental protection results in short-term cost advantages for a few, and enormous long-term cost liabilities for many. That is why the United States needed a Superfund to pay for the clean-up of the worst land pollution sites in America. New Jersey, as one might guess, has more Superfund sites than any other state. While the Superfund site list is not the list New Jersey wanted to be number one on, it is at least an indication of the work in New Jersey today to clean up past land pollution sins. In 1994, New Jersey ranked fifth of the 50 states with $236 per capita for environmental spending by state and local governments. New Jersey may have been left with some of the worst environmental problems in the nation but we were trying to clean them up. Even though our ocean clean-up project was impacted by budget cuts, New Jersey was still among the big spenders on environmental issues.

THE DUTCH GREEN PLAN

In her 1993 election campaign, Governor Whitman committed to reduce taxes and government costs, which she did. However, having come from a family that owned a 158-acre farm in Tewksbury, New Jersey, she also was interested in improving the environment. In 1994, Governor Whitman and the DEP's Commissioner Shinn visited the Netherlands at the invitation of the Dutch government and the New Jersey Conservation Foundation to learn about the Netherlands National Environmental Policy Plan (NEPP). The focus of the trip was to understand how the Netherlands had developed a long-term plan to both protect the environment and ensure a sound economy. Our new governor wanted to accomplish both economic and environmental objectives, because she knew that one could help the other.

A Benchmarking Trip

This type of trip, focusing on learning how to improve your key processes, is referred to as a benchmarking trip in our quality approach. Benchmarking, to help establish goals and measures, and the deployment of goals are the key steps to the "plan" phase of the plan-do-check-act (PDCA) cycle of quality improvement. While the governor and DEP commissioner did not understand all the PDCA work we had been doing on the New Jersey shore water quality, they were taking part in the first step of the PDCA cycle required for continuous improvement of the entire New Jersey environment.

I learned about this trip in 1995 when I was invited to represent Quality New Jersey at a conference held to develop measurable goals for sustainable growth in New Jersey. The conference was hosted by New Jersey Future, a nonprofit organization aimed at bringing together government, business, and environment groups to provide input to a plan for sustainable growth in New Jersey that would support growth while protecting the environment. The conference was held at Princeton University and about 70 individuals attended along with the governor and the DEP commissioner.

The conference started with a sharing of the Netherlands' benchmarking trip information. The Netherlands and New Jersey are similar in many ways, which made the Dutch plan of particular interest. Some of the similarities are as follows:

1. The high population density of New Jersey, highest of any state in the United States, was virtually identical to that of the Netherlands.
2. New Jersey leads the United States in the number of Superfund sites and the Netherlands leads Western Europe in terms of soil contamination.
3. Both places have serious air and water contamination.
4. Holland is a port country and New Jersey is a port state.
5. In both places, the chemical industry sector is the largest business sector in the economy.
6. The country of Holland and the state of New Jersey share similar economic traits.

The key to getting started on their environmental plan in the Netherlands was the release of a report with measures of the problem by the Netherlands National Institute of Public Health and Environmental Protection in 1988. The document, entitled "Concern for Tomorrow," revealed a number of serious environmental problems, such as the following :

1. More than 50 percent of the country's trees were damaged by acid rain.
2. Inland waterways contained 10 to 25 times more nitrates and phosphates than would occur naturally, causing algae growth and disruption to the aquatic life.
3. Sixteen hundred instances of soil pollution had been identified as needing treatment.

A Call for Action

Queen Beatrix of the Netherlands saw the report and delivered a speech that was a call for action. In response, Prime Minister Lubbers issued a National Environmental Policy Plan (NEPP). Four major policies were contained in the NEPP:

1. Not to increase the level of pollution in the air and water nor create a greater degree of pollution in any place while solving a problem elsewhere.
2. To remediate air, water, and soil pollution to a health standard stringent enough to prevent pollution-related disease by the

year 2010, and to allow human and wildlife use of the air and water without detriment.

3. To reach an ambient level of air and water contamination that could meet all health standards through a negotiated consensus process with all segments of society.

4. To establish a firm policy that the polluter should pay for remediation.

Adoption of quantified targets and timeframes is central to the plan, which also is based on the premise that controlling environmental problems at the source is the most efficient means to reduce pollution. Municipalities are under no obligation to follow suit but receive additional central government funding if they do. Financial support motivates use of best practices. Agreements with industry have been forged that essentially consist of voluntary commitments to decrease emissions and discharges. Industry is free to determine how it will reach specific agreed-to goals.

In March 1993, a new Environmental Management Act radically simplified Dutch environmental law. Numerous old laws protecting nature and landscape and regulating emissions were replaced by a single statute. In addition, a single point of contact and management was assigned in their environment enforcement group for each applicant, coordinating all permits for each applicant.

There is a distinct separation between developed and undeveloped land in Holland. Town is town and country is country. Country areas are defined and protected from development. Developers are required to construct in areas so zoned. Consequently, despite its equally high population density, Holland does not have the unguided sprawl that characterizes New Jersey. In Holland, 65 percent of Dutch land is still in agricultural use compared to New Jersey where only 16 percent is still agricultural.

In Holland, developers are allowed to build only after the local government has constructed the required infrastructure in accord with national standards. The infrastructure includes roads, storm drains, sewers, and other necessary utilities.

Holland also earmarked $5 billion over a five-year period to support their new NEPP policies, which provided the financial resources necessary to support their goals.

NEW JERSEY GREEN PLAN DEVELOPMENT

In New Jersey, at our Princeton conference, it was acknowledged that we did not have a goal-setting system and that we had not done a good job of monitoring the effectiveness of our laws over the years.

The benchmark in Holland was an encouraging example of movement from a focus on micro-regulation and confrontation without goals, to macro-regulation and cooperation with goals.

The conference at Princeton suggested that the role of government was not to solve all of society's problems, but to help business and citizens to become active participants in working toward solutions to the problems in society. This sounded like what our QNJ environment team had been trying to do since 1989 on our Jersey ocean water quality improvement effort—and this approach worked here in New Jersey.

A number of New Jersey environment organizations had also gotten together to develop a "Green Plan for New Jersey" as input to the governor's efforts to develop a New Jersey environment plan. While this initial plan included a number of principles, guidelines, and strategies it lacked the quantifiable measures and goals that were contained in the report by the Netherlands National Institute of Public Health and Environmental Protection report entitled "Concern for Tomorrow."

The conference planners, in advance of the conference, had assembled some selected potential measures and data on the quality of the environment in New Jersey that might be used in developing the New Jersey Green Plan.

Water

The annual New Jersey DEP Office of Land and Water Planning report, "New Jersey 1992 State Water Quality Inventory Report," provided detailed data about water quality in New Jersey's 6,400 miles of rivers and streams, 24,000 acres of public lakes, 90,000 acres of freshwater and tidal wetlands, 120 miles of open coastline, and 420 square miles of open estuary waters.[4]

1. Of the 525 stream miles monitored for primary contact (swimming) use, 85 percent did not fully meet the standards.

2. Of the 1,421 stream miles about which there was information, 27 percent did not support aquatic life.
3. In 1990, 430 square miles of coastal waters and 614 miles of estuary and bay waters were assessed for fitness for shellfish harvesting and 27 percent and 28 percent, respectively, were not fit for shellfish harvesting.

In summary, 85 percent of our New Jersey streams were not fit to swim in and more than 25 percent of our streams, ocean, and bay waters did not support healthy aquatic life in 1990. Not a good situation.

Our QNJ environment team had focused primarily on improving the ability of the New Jersey shore beaches to be open for swimming, which was not one of the key environmental measures initially selected by the conference planners. Given the New Jersey voters' high priority on having the Jersey shore ocean beaches open to swimming, I suggested a measure to be included in the proposed New Jersey Environmental Green Plan.

If the Jersey coast has about 500 blocks of good beaches, then in a 100-day summer season we would have 50,000 potential beach-block-days of swimming. If we had 500 beach-block-days of closures during the summer it would be 1 percent of the swimming opportunities, which sounds like it is not that bad. However, we knew that 500 beach-block-days of closures during the summer was unacceptable to the New Jersey beachgoers. Therefore, because anything greater than zero was less than desirable, rather than use the percentage closed to swimming, I suggested that the number of actual beach-block-days closed due to pollution be used as the measure.

Air

The New Jersey DEP Bureau of Air Monitoring "1993 Air Quality Report" reported on conditions and trends for the past decade.

According to the Pollutant Standard Index (PSI) air quality in New Jersey in 1993 was good for only 66 days (18 percent), moderate for 280 days (77 percent) and unhealthful for 19 days (5 percent). Not a good situation. Of the unhealthful days, 18 of the 19 were due to ground level ozone—a major component of smog.[5]

Land

As of 1990, approximately 2.8 of New Jersey's 4.8 million acres were developed. New Jersey has lost half its farmland since 1950. In 1990, 870,000 acres (18 percent of New Jersey) remained as farmland. As of 1995, there were 779,891 acres (16 percent of New Jersey) of public open space reserved for public use and the preservation of natural resources.

Since the inception of the Superfund in the early 1980s only seven of 114 Superfund sites had been completely remediated, while others had been partially remediated.[6] In addition, there were 21,000 other hazardous discharge sites of which 6,076 (29 percent) were known contaminated sites. In 1987, the list for the nation contained some 964 Superfund sites, with New Jersey on the top of the list, followed by Michigan, California, Pennsylvania, and New York.[7]

New Jersey was well beyond the environmental crisis that the Dutch had responded to in Holland. New Jersey had 85 percent of its streams not fit for swimming, 82 percent of its days not good for breathing, the most Superfund sites in the nation, and an additional 21,000 other hazardous discharge sites. New Jersey needed a response from the New Jersey leadership similar to what the Dutch had mustered, with measures, goals, and financial support for improvement.

In 1984, the state had passed the New Jersey Water Pollution Control Act, which had to conform with the federal Clean Water Act. Yet, by 1995, we still had not agreed on how to implement a policy of not degrading our water.

The conference did not have a follow-up meeting in 1996, but on one of my visits to the New Jersey DEP in 1996, I was given a glimpse of an environmental plan, developed by the regional EPA and New Jersey DEP, which established goals, measures and plans of action. This was the first such joint federal and state government plan ever produced for New Jersey. While this initial environmental plan for New Jersey would not be satisfactory to everyone, it was a start. In addition, there was a commitment on the part of the EPA and DEP to focus their resources toward meeting the goals in the plan and to update and improve the plan annually. Of course, public sharing of the goals and the plan by the New Jersey DEP and the EPA, as was done in Holland, would probably have been a good step to improvement of the plan and to improvement of the resources made available to support the goals in the plan.

The DEP unveiled its plan for reworking of the state's water quality programs to implement the anti-degradation policy in the fall of 1996. It was initially rejected by the environmental community. This initial plan addressed the New Jersey Pollution Discharge Elimination System (NJPDES), water quality standards, and water enforcement. The DEP went back to planning and released a draft statewide plan that addressed only the New Jersey Discharge Elimination System (NJPDES) for watershed management, without addressing land use.

To protect the New Jersey drinking water supply, in 1988 the New Jersey legislature had passed the Watershed Moratorium Act. This act put a halt to the transfer or sale of any water-utility land pending the development of some regulations and buffers by the New Jersey DEP. In addition, a Watershed Review Board was created that could grant waivers based on very narrow criteria. At that time, the review board was headed by the president of the state's Board of Public Utilities, who was Christine Todd Whitman, the DEP commissioner, and the head of the Department of Community Affairs.

In 1997, this moratorium was still in effect because the DEP had not developed new regulations and buffer requirements to protect the New Jersey drinking water system. Perhaps the best protection was simply to leave this existing moratorium and review board in place. Owners of the land affected by the moratorium are protected against having towns charge them higher property taxes as long as the moratorium exists, which is a positive incentive to encourage the landowners to go along with the moratorium.

For a long time there was a belief that the technology exists to remove whatever we put into the water. Even if this were true, technology can have a very high cost, which results in it being less than fully deployed. Preventing pollution at the source is the most efficient and effective means of avoiding pollution, not treating it at the end of the line. One example of prevention is having a moratorium on building on buffer land near a water supply.

Another example would be to invest in rebuilding our existing towns and cities instead of building on new land in New Jersey, given the state of urban sprawl that already exists. We have a tendency to let a city run down and then simply move on to building and buying property in a new town or city. In the Netherlands, developers are not allowed to build on "country" land. Consequently, citizens must take care of their community and invest in it to ensure that they have a nice place to live. This also provides a degree of land, water, and air protection, in the "country," both quality and quantity for all. Redevelopment also increases values for current property holders, thereby providing an additional incentive for improvements and expansion on existing property.

Providing a land buffer around state water supplies provides a degree of protection for the quality, and quantity, of water required by the state. Polluted water supplies reduce the quantity of clean water available. In addition, the feeder streams and tributaries must also be protected. To do this, a watershed management program is required that also protects the land around the feeder streams and tributaries.

AT&T'S GREEN PLAN

While the state struggled to establish measures and goals in 1995, AT&T was reviewing how it did on the goals it had set in 1990 for its own AT&T environment plans. AT&T was also developing new goals for the year 2000.

1. In 1990, AT&T set a goal of eliminating the use of ozone-damaging CFCs by the end of 1994. The company achieved this goal in early 1993.
2. In 1990, AT&T pledged to reduce reportable air emissions 95 percent from the level in 1987, by year-end 1995. By May 1995 they had been reduced by 96 percent.
3. In 1990, AT&T set a goal to reduce manufacturing waste by 25 percent from the 1987 level. AT&T managed to decrease it by 66 percent.
4. In 1990, AT&T set a goal to recycle 60 percent of its paper by the end of 1994. AT&T achieved a 65 percent recycling level. The company also reduced its use of paper by 29 percent.

AT&T committed to using quality policy deployment and a quality approach to achieve these goals and the new goals being set for the year 2000.

One of the new programs AT&T began to emphasize in 1995 was telecommuting for both environmental and productivity reasons. Alvin Toffler, author of *Future Shock* and *The Third Wave*, has said, "Commuting is the single most anti-productive thing we do." Long hours in traffic is time poorly spent. Telecommuting is a fancy term for working at home while staying in touch with the office through computers, e-mail, Internet, faxes, and phones. In 1995, nearly 30,000 AT&T employees telecommuted part-time and about 5000 worked from home full-time. AT&T also supports the "Telecommute America" week in October when employers and employees nationwide are encouraged to try staying home to work.

It is a fundamental fact of life that driving produces carbon monoxide. Improving air quality is another important reason to encourage telecommuting. Of course, AT&T is in the business of providing telecommunications, which is perhaps an additional reason for AT&T to encourage and support increased telecommuting.

QNJ ENVIRONMENT TEAM PLAN

While the nation, the EPA, the DEP, and AT&T were all working on their goals, our QNJ environment team had to get back to work on our own Jersey shore water goals. Our team scheduled a picnic meeting on May 23, 1995 at Sandy Hook, the northernmost beach in New Jersey that connects to the New York–New Jersey estuary.

The information I received at the Environmental Sustainability Conference in Princeton was shared with my QNJ environment team. I also shared my disappointment that our work on cleaning up the ocean and bay water quality was not very visible to others, since one of the purposes of our work was to create a model for the use of a quality approach in the environment, and in government in general, that others might replicate. The group discussed possible ways of making our work more visible to encourage similar work in other environmental areas in New Jersey. Dave

Rosenblatt, the DEP ocean water-quality monitoring supervisor, suggested that the best hope for our approach being used in other areas would be to work with Nick Finamore, an AT&T executive who had recently been loaned full time to the New Jersey DEP to help the DEP launch its quality program, "Innovations through Quality." Because I knew Nick, I agreed to contact him to share the work of our team and offer quality support to him in his new loaned executive role.

Peter Brandt, the regional EPA representative on our team, indicated there had been no closings of New Jersey beaches over the last four years due to floatables and no closures of New York beaches due to floatables for the last two years, thanks in part to the work of the regional EPA office. Peter also suggested that Jeanne Fox, the New York EPA regional director, attend the 1995 QNJ conference and participate in the presentation of the Shore Quality Awards to the best shore towns, along with the New Jersey DEP commissioner and me. We had come a long way from 1992, the initial year of the award, when the DEP staff wouldn't let the DEP commissioner participate in the awards presentation. Peter also shared news on the proposal in the Republican-controlled House of Representatives in Washington to cut the EPA program funding by one-third in 1996.

Dave reported that his state-level water monitoring department had a staff of four, compared to two years before when he had a staff of eleven. He was learning to do more with less using cooperative programs involving the shore counties and municipalities. Dave reminded us that the four coastal health departments were responsible for actually taking the weekly water quality measures, and that his staff collects, reports and analyzes data for the entire state.

For the 127-mile New Jersey coastline we have 180 water-quality monitoring stations on the ocean side and 123 monitoring stations on the bay side. On the ocean side we also have 150 storm-water outfall pipes and 17 sewage-plant outfall pipes, which are all potential sources of pollution. On the bay side we had 7,000 storm-drain outfall pipes, each a potential source of pollution, but no sewage-plant outfall pipes. New Jersey had at least agreed that putting a sewage outfall pipe into a bay was not acceptable, unlike some other states, like Massachusetts, that continued that practice in 1995.

Our 1994 beach-block-day closures due to pollution had increased in part due to the 1994 budget cut and the pressure on reducing costs, which continued in 1995 at both the DEP and EPA levels. The summer season was to begin soon with Memorial Day, and we needed improvement in spite of past budget cuts.

Laurie Groves, one of our key team members who did the bulk of the communications support for the team, reported she had sent the application for the QNJ Shore Quality Award, along with the completed 1994 application from Avalon as a municipal role-model for other towns to replicate, to all 90 shore municipality mayors as well as the four shore county health officers. In addition, she had sent this to the 34 New Jersey shore local newspapers with a suggested press release to encourage more shore interest. The majority of the work to improve our shore water quality had to be done by the shore counties and the shore towns, not the state DEP or federal EPA. In most cases their revenues were increasing due to summer tourism and they could afford to make improvements that protected the shore water quality, which was needed to generate their tourism revenue.

Dave reported that the New Jersey shore license plate (lighthouse picture and motto "Shore to Please") program sold 39,000 plates in 1994, which generated about $1 million of dedicated citizen-contributed funds, to replace DEP budget cuts, for four of our state-level shore clean-up programs:

1. Operation Clean Shore (prisoners used to clean up beaches)
2. Pump Out Facilities (upkeep of facilities)
3. Shore Surveillance (salaries for staff)
4. Adopt a Beach (state support of town efforts)

Using Avalon as a municipal role model, license plate funding from supportive citizens for our state programs, and continuing our quality award process to increase awareness through recognition, we planned to get back on track toward our goal of zero beach closings. We knew the problems, we knew the root causes, we knew the countermeasures required. We continued to face the difficult job of deployment. Deployment requires motivation and resources. Our team was doing its best to provide support for both, in spite of state- and federal-level budget cuts.

While at Sandy Hook, we also visited the National Oceano-graphic and Atmospheric Agency (NOAA) laboratories. The NOAA was in the process of building a new laboratory for students and researchers to study the behavior of various kinds of fish. The laboratory had a plumbing system that enabled them to run ocean water into the laboratory tanks to duplicate the various environmental conditions in which the fish actually live. The tanks were equipped with cameras that run 24 hours a day for the scientists to study the next day. On a beautiful day at the beach, our plan for a picnic ended up as a quick lunch in a meeting room so that we all could fit in the NOAA lab tour.

NEW JERSEY TOURISM

Unfortunately, I didn't get a chance to go back to the Jersey shore for vacation in 1995 because I chose instead to invite my kids for a two-week vacation on a lake in Greenfield, New Hampshire, at what had been Nana's and Papa's cottage. Nana and Papa had passed away and the cottage had been sold, but I rented it for two weeks so that my kids and my two new grandchildren could have a chance to go to Nana's and Papa's lake house, a place they had loved as kids.

In 1988, when I dragged my four kids off to the Jersey shore for a vacation at a beach that was closed to pollution, they ranged in age from 11 to 19 years old. In the summer of 1995 they now ranged from 18 to 26, and I had two grandsons, ages 1 and 3. I thought they all might like to return to what had been their favorite vacation spot as kids. In spite of my vacation-planning track record, they all came and they all enjoyed the opportunity to remember the good times at the lake with Nana and Papa.

At the end of the two weeks, John and Michelle, my youngest, said, "Thanks for the vacation at Nana's and Papa's lake, but without them here we would prefer to go to the Jersey shore next year."

Around the Fourth of July, a New Jersey *Star Ledger*–Eagleton Poll was done to see how New Jersey citizens felt about New Jersey as a vacation spot. More than 70 percent rated New Jersey good or excellent, 22 percent rated New Jersey fair, and only 5 percent rated the state as poor. In addition, 80 percent indicated they intended to visit the Jersey shore in 1995.

When New Jersey citizens were asked to name their favorite attractions, the results put the Jersey shore at the top of the list, and leading by a wide margin:

1.	Jersey shore	40%
2.	Atlantic City's casinos	10%
3.	Parks	7%
4.	Movies	4%
5.	Great Adventure Park	3%
6.	The Meadowlands	2%
7.	The Garden State Arts Center	2%
8.	The Liberty State Science Park	1%
9.	The new aquarium	1%
10.	The Delaware Water Gap	1%
11.	New Jersey rivers and lakes	1%
12.	Restaurants	1%

This survey provided a good indication that the New Jersey beachgoers were feeling a lot better about the shore than when we started our improvement team efforts in 1989.

STRETCH GOALS

With all these people liking the shore and wanting to go there, we were relieved to see the 1995 season turn out to be our best ever for shore water quality. The total number of beach closings on the ocean and bay was 77 compared to 855 in 1988. We had achieved more than "a factor of 10 improvement" since 1988. Use of a quality approach could achieve a factor of 10 improvement in all our key New Jersey environmental measures. The deployment of a New Jersey Green Plan would require a quality team using a quality approach for every key area.

New Jersey now finds itself in an environmental crisis situation, as a result of many years of neglecting care for our infrastructure and environmental issues. In seven years we had reduced the shore water quality problem to less than 10 percent of what it was in 1988. Now that we had shown it could be done, a goal for other areas could be to make a similar factor of ten improvement in seven years on key environmental measures. "Ten in Seven" could be the Jersey environmental goal. While this is a stretch goal, achievable stretch goals are required to motivate the significant change in behavior that is required to obtain significant improvement. Stretch goals also call for the appropriate financial support, which could be provided through bonds, which the public probably would support if funds were to be dedicated for the purpose of achieving the stretch goals established for improvement of the environment in New Jersey. It took many years of neglect to create the environmental crisis in New Jersey; it cannot be turned around using a bit of expense from one year's budget. A long-term investment is required, which the public probably would be willing to finance, if dedicated to reaching stretch goals for environmental improvement in New Jersey.

BAY BEACHES

On ocean beaches we had only four beach-block-day closures in 1995, compared to over 600 in 1988, coming close to achieving our original goal of zero on the ocean beaches. On the bay beaches we had 73 beach-block-days of closures, which was a significant improvement, more than a factor of three over the 1989 results, but the bay water quality was improving at a slower rate than the ocean water quality (Figure 8-1).

Recognizing the different rates of improvement, we analyzed the reasons for the better rate of improvement on the ocean beaches. The primary sources of the ocean pollution had been combined pipe overflows of raw sewage, floatables, sewage treatment plant outfall pipes and the run-off from the 150 storm-drain pipe outfalls. The sewage treatment plant problems had been successfully eliminated, the floatables had been reduced, and the repair of storm-water sewage pipe problems had been improved. In addition, by 1997 we had obtained a commitment on the part of New Jersey to spend more than $1.3 billion to deal with some of the problems from combined sewer overflows into the Hudson River estuary.[8] Another big help to improvement on the ocean beaches was the economic motivation of shore towns for improve-

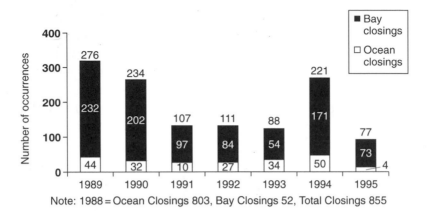

Note: 1988 = Ocean Closings 803, Bay Closings 52, Total Closings 855

**Figure 8-1. New Jersey Beach-Block-Closings Trend—
Ocean and Bay Results**

ment because beachgoers who spent the day at the ocean wanted to be able to go in the water.

On the bay beaches, we had 7000 storm-drain pipes and less natural filtration to disperse any pollution leaking into the bay. Barnegat Bay, in Ocean County, is a 40-mile thread of water between the mainland of New Jersey and Long Beach Island. The bay is only four miles across at its widest point and has an average depth of about five feet. Its waters heat quickly and stagnate easily, primarily because replenishing ocean waters can enter only through two narrow inlets on its eastern coast. The tides on the ocean side were an automatic means of taking a pollution problem out to sea—thanks to the moon, not our team. In addition, the bays were seen as a place for boaters, not swimmers, which reduced the motivation for improvement in the bay beach water quality.

Potentially the worst of the bay's troubles, according to Pete McLain, a wildlife biologist and former deputy director of the New Jersey Division of Fish and Wildlife, is a single-cell protozoan he discovered is killing the bay's beds of slender pale-green reeds, called eel grass. The beds serve as critical breeding grounds and shelters for juvenile fish, clams, and blue crabs. Any widespread loss of the beds will seriously deplete those stocks, according to McLain.[9]

Our team wanted to improve water quality faster in the bays. We wanted parents to be able to swim there with small children, avoiding the dangerous pull of the currents on the ocean side. We had 12 drownings on the ocean side in 1995 due to rough water, after-hours surfers, and the undertow caused by the jetties that were put in place many years ago by the Army Corps of Engineers in an early attempt to protect the sand on the northern shore ocean beaches. The jetties didn't work, but the Army Corps of Engineers has not been directed, or funded, to remove them.

A QNJ Shore Quality Award winner from 1992, Stafford Township, had a lot of bay beaches. To protect their bay beaches, Stafford Township had created a natural aquifer (a group of rock formations) and terminated their storm-drain pipes in it—rather than into the bay. If you can't get all the pollution out of the pipes, at least you can get the pipes out of the bay. The aquifer provides a natural filter for the water in the storm-drain pipes, before it ends up in the bay. The bays had some different root causes than the ocean for their pollution and therefore needed some different countermeasures to deal with the causes.

Our team needed to share more than one role model. For the ocean beaches Avalon was a role model in many ways. For the bay beaches, Stafford Township was a role model. For improving shore sewage treatment plants, Asbury Park had become a role model.

SUSTAINING THE GAIN

Continuous improvement requires continuous discovery of root causes of problems and continuous development of new counter-measures to eliminate them. It also requires deployment and continuous maintenance of countermeasures for the root causes previously identified. After several years of working on the shore water quality problem our team and the shore counties and towns had identified and developed a countermeasure for most of the initial root causes of the shore problems. We had also made good progress in having the shore towns deploy and maintain those countermeasures. However, continued monitoring, recognition, sharing of best practices, resource support for required improve-ments, and analysis of new root causes was required to maintain

the gain and continue toward our goal of zero beach-block-days of closures on both ocean and bay beaches.

A lack of support for continued implementation of the necessary countermeasures, repair of the dam holding back pollution, would simply result in the return of the original root causes, which could return the Jersey shore to the worst in the nation. When a quality goal is reached there is the danger of thinking the problem is solved. Instead, it should be understood that countermeasures have been put in place to address the root causes of problems and the causes will return if we stop supporting the countermeasures. Reducing the required support for the countermeasures that are preventing the root cause of a problem, to go on to solve another problem, results in the return of the original root causes. Discovery and development is more fun, but maintenance is also necessary to sustain improvement. For example, once the ocean water quality seems to be good, in order to save that monitoring cost some people feel we no longer need to monitor the quality, because we currently do not have a problem. They do not realize that we do not have a problem because we are monitoring the water and using the results to drive fast repair of pollution problems. However, in the case of monitoring, as the quality gets better, the sampling interval could be less frequent. If we had monitored once per day when we had a serious level of pollution, then as the water got better we could have reduced the sample to once per week, which we were now doing.

House Bill to Lower Our Water Standards

In 1995, we were making progress on our Jersey shore water improvement project, New Jersey was beginning to develop a statewide environment improvement plan, and AT&T was setting environment goals for the year 2000. However, at the federal level, we were also facing potential serious water quality setbacks.

First the House passed the Clean Water Amendment Act of 1995 (HR961), which allowed the municipalities and counties throughout the nation to delay fixing their sewage and storm-drain sources of pollution until the year 2010. The reason for this bill was to postpone the cost to the municipalities and counties of fixing their infrastructure, while they continued to pollute our waterways.

Protecting Future Generations

When I was 18, I wrote my senator in Massachusetts to ask for the reduction of the voting age from 21 to 18, since at 18 a person could be working and paying taxes, or in the army fighting for our country. It seemed to me that when you were old enough to pay taxes, and die for your country, you should have a say in how it is run. When I was 22, I received a letter back from my senator, Ted Kennedy, thanking me for my earlier letter and informing me that he had gotten a bill passed lowering the voting age to 18. This action, and response from Senator Kennedy many years later acknowledging my input, left me with a feeling that citizen input for improvements in our governance process were considered, and could make a difference.

However, there now seemed to be an inclination in the House to push pollution onto the next generation, a generation that isn't even 18 yet, and doesn't have a vote in what may be pushed on them. While kids do not yet pay taxes or fight for their country, decisions are being made today that will affect the taxes they have to pay in the future and affect the environment they will have to live in. I think they should have a say in those decisions that affect their taxes and their environment. The Constitution states: "We the people of the United States . . . promote the general welfare, and secure the blessings of liberty to ourselves and our posterity." Our posterity is the next generation, which our Constitution demands we look out for. When we push pollution and costs onto the next generation instead of dealing with them ourselves, we are violating a key value in our Constitution. This is the type of value our Congress and our Supreme Court should be protecting, not violating.

The postponement of the infrastructure costs would be to the next generation. In addition to leaving them the cost of improvements, we would also leave them 15 more years of pollution to clean up.

The Clean Water Act of 1972 recognized the need to develop plans to eliminate municipality and county dumping of raw sewage into our rivers and waterways. In 1995 we still had thousands of combined sewer overflow points in the United States that are sources of raw sewage pollution whenever a heavy rain occurs. Due to the increased flow of water in the combined pipe due to the heavy rain, raw sewage overflows directly into our waterways. In 1995 we still had no plans to replace these combined pipes, which were a major source of pollution. The reason for this bill being introduced and passed in 1995 was that the EPA had finally begun pushing counties and municipalities to comply with the 1972 law and some counties and municipalities had run to their congressmen to get the EPA off their back. In the name of supporting their local state constituents, this bill was introduced and passed in the House. The House of Representatives was putting their current state interests ahead of the nation's best interest or the best interests of future generations. This same thinking was what led to a House bill to cut the EPA enforcement budget by one-third in 1995.

One solution to the combined sewer overflow (CSO) problem is to build treatment plant capacity that handles the peak volume of both heavy rain and raw sewage flow. This has been done by a least one affluent county in California. Another solution could be simply to put in the separate pipes required for the storm-drain and sewage, which should have been done in the first place. This simply requires digging a trench and laying a new pipe.

Fortunately, with letters from many opposing this turning back of the Clean Water Act, the Senate did not support the House bill and therefore it did not become law in 1995. However, it does provide an indication of the lack of understanding and support for environment issues by the majority in the House of Representatives.

House Bill to Eliminate the National Quality Award

The budget debates in Congress in 1995 resulted in proposed cuts for many things, in addition to the EPA cuts. One proposal was for the elimination of the Department of Commerce and with it the Malcolm Baldrige National Quality Award. In 1995 Bob Allen had just been elected president of the Malcolm Baldrige Foundation, a voluntary position to ensure that the National Quality Award received the appropriate financial support from the private sector for this public–private partnership. My office ended up developing material for a congressional hearing on the value of continuing the National Quality Award. The material was presented by one of the foundation's members and the congressional committee agreed that this was a great partnership that returned tremendous results for minimal costs and should be continued. In spite of the committee's support, the House passed a bill that eliminated funding for the Malcolm Baldrige National Quality Award.

The lack of understanding within the House of Representatives of the benefit of encouraging a quality approach to improve the economy also became apparent. Again, with the support of many who wrote letters about this most beneficial and minimal cost program, the Senate supported continuing the award.

TELECOMMUNICATIONS ACT AND AT&T SPLIT-UP—AGAIN

As the year ended in Washington with budget debates, the debate on a new telecommunications bill to open the U.S. local telecommunications market to competition, as had been done in the long-distance telecommunications market, was pushed into early 1996. The purpose of this bill was to open the monopoly local market to competition. However, the monopoly local telephone companies were, of course, very vocal in Washington about what they wanted. As a large, state special-interest group, they were also successful in getting the support of the House leadership for their wishes.

The competitive long-distance companies, such as AT&T, MCI and Sprint are all required to pay excess access charges to the monopoly local telephone companies simply to be connected to the local monopoly network. The local monopoly telephone companies convinced Congress not to require the explicit elimination of these excess access charges in the new Telecommunications Act, even though they were not cost-justified and the long-distance companies could reduce their rates by about 40 percent if these local monopoly excess access charges were eliminated.

The Telecommunications Act instructed the FCC to ensure that implicit subsidies (which referred to these excess access charges) were eliminated when they implemented the regulations to enforce this new Telecommunications Act. However, in 1996, the local telephone companies went to court to block the FCC regulations on the grounds of states' rights in setting local prices. This was a tactic to maintain their local monopoly excess access prices. Whether at the federal, state, or local level, the few who have the most to gain have the opportunity to be more vocal than the many who each have a small amount to lose. In some cases, it may be the quality of our water, in other cases it may be the price we pay for our monthly telecommunications bill.

On September 20, 1995, in anticipation of the 1996 Telecommunications Act, Bob Allen announced that AT&T would split up into four companies: AT&T for telecommunications services, Lucent Technologies for telecommunications manufacturing, NCR for computers, and Capital Corporation for financing of telecommunications purchases. The existing AT&T corporate functions would be split and go either to AT&T, Lucent, or NCR. Leaders of the new companies, who had been or were being named, would decide how large of a staff they wanted and who they wanted to take from the existing AT&T corporate staff.

This decision to split up AT&T was required by the impending Telecommunications Act, because after the act was implemented, the AT&T telecommunications services units would be competing with the major customers of the AT&T telecommunications manufacturing units.

Transition Planning

The corporate quality office staff of 26 was told that we would not continue as a team and no one, including me, could be sure we would have a job in one of the new companies in 1996. The existing AT&T corporate quality office would be phased out by April 1, 1996. The entire office began preparing their résumés and developing an alternative work plan, should we not be selected to go to the quality office of one of the new companies.

During this difficult transition period, the team helped each other look for the jobs we each wanted, and we worked together to develop a transition of the past work of the old AT&T quality office capabilities to the new companies.

1995 QNJ QUALITY CONFERENCE

In spite of our AT&T transition work and job uncertainty, the QNJ environment team was able to select and recognize four shore municipalities for their excellent work on shore water quality at our annual QNJ quality conference on October 31, 1995. A sign of improvement in the government support for our effort was the increased participation of key individuals in the presentation of the 1995 QNJ Shore Quality Awards. New Jersey commerce commissioner Gualberto Medina, New Jersey DEP commissioner Robert Shinn, and the New York–New Jersey area EPA administrator, Jeanne Fox, all joined me in recognizing the accomplishments of our winning shore municipalities: Avalon, Asbury Park, Brigantine, and Manasquan were the QNJ 1995 shore municipality award winners, all with zero beach closings.

- Avalon had become an annual winner, and continued to demonstrate improvement.
- Asbury Park had made a turnaround since their treatment plant failure in 1988. For a municipality still struggling financially they had made great efforts to ensure they were preventing shore pollution.
- Brigantine had replaced deteriorated storm-drain and sewage pipes and implemented a number of trash containment programs.
- Manasquan, like Avalon in years before, had focused on reducing water infiltration into the sewage system, which also reduces sewage infiltration into the storm-water system. The result is lower treatment costs and cleaner water.

At the conference, we played a two-minute video that was developed by the Juran Institute (a major quality consulting organization) on our QNJ team's work in using a quality approach to solve an environmental problem. This video was shared with all the Juran Institute customers as part of a series of short videos produced by Juran on quality projects that others may be able to learn from. The CEO of the Juran Institute, Blan Godfrey, asked me if I would be willing to share the story of our team for one of his two-minute videos. I was happy to help because that was the purpose of the project in the first place, and because Blan was an old friend who had worked with me when he had been a department head at our AT&T Bell Labs quality center.

Our 1995 shore quality results were the best ever. We had improved by more than a factor of 10 since 1988 (855 beach-block-day closings in 1988 to 77 beach-block-day closings in 1995) and we had only 4 beach-block-day closings on ocean beaches in 1995.

Our New Jersey shore tourism continued to rise at about $1 billion per year, putting us behind California and Florida, and ahead of Hawaii in shore tourism revenue.

CALIFORNIA AND FLORIDA— SHORE WATER QUALITY PROBLEMS

California

While New Jersey had improved by a factor of 10, California had gotten worse by a factor of 10. According to the NRDC July 1996 report on 1995 shore water quality results, California reported 1305 beach closing days and also had 11 permanent and 3 extended beach closings. In addition, these results were only from the nine counties, of the seventeen shore counties in California, who regularly monitor and report on their water quality.

Five of the nine counties that monitored their shore water quality in California each had closings or advisories that were higher in number than the entire state of New Jersey in 1995:

1. Orange—132
2. San Mateo—180, plus 1 extended
3. Santa Barbara—224, plus 3 permanent
4. San Francisco—331
5. San Diego—338, plus 3 permanent

California was having problems similar to those that New Jersey had been dealing with in 1987 and 1988, and needed similar solutions. San Diego beaches were impacted by sewage-contaminated water from Mexico's Tijuana River, which discharges at southern San Diego beaches. Due to a lack of treatment-plant capacity, Tijuana was diverting an average of three million gallons of raw sewage into river and ocean waters every day. The Mexican government and our EPA joined forces to construct a new sewage treatment plant that was expected to be on-line by 1997.

Storm-water pollution is the largest source of pollution to San Francisco Bay and Santa Monica Bay. The Santa Monica Bay Restoration Project recently completed a study of the health effects of swimming in Santa Monica Bay. The study found an increase in risk of colds, fever, chills, sore throats, diarrhea, and other symptoms of illness in those swimming near storm drains, as compared to those swimming farther away. Raw sewage overflows were the main cause of pollution in Orange County.

Florida

In Florida, only 11 of the 34 coastal counties with swimming beaches conducted monitoring for swimmer safety in 1995. Even with only one-third of the counties doing monitoring, Florida had 830 beach closings and advisories in 1995. Again, sewer systems and storm drains were the primary sources of the pollution, along with hurricane Opal, which flooded septic systems and caused debris.

In 1995 California and Florida had the most reported beach closings and advisories per the annual NRDC report.

Neighbors to the North

Following California and Florida in the number of beach closing and advisories in 1995 were our neighbors to the north:

1. New York—283, plus 3 extended
2. Connecticut—252, plus 1 extended
3. Massachusetts—132, plus 1 permanent

New York City has a significant problem from combined storm-drain and sewage-system pipes. According to the *New York Harbor Water Quality Survey* of 1993, 70 to 80 percent of New York City's 6000 miles of sewer system are combined with storm-water pipes, which can discharge a mixture of rainfall runoff and raw sewage into area waterways during and immediately after precipitation. Governor Whitman stated that 78 percent of the uncontrolled discharges of raw sewage into the Hudson River estuary were from New York, with 22 percent coming from New Jersey.[10]

In 1995, an advisory press release was issued just before Memorial Day weekend warning against swimming at Bronx private beaches 48 hours following a rainfall of 0.2 inches in 2 hours, and against swimming at South and Midland beaches on Staten Island and Manhattan Beach in Brooklyn for 12 hours following rains greater than 1.5 inches. In other words, whenever it rains we get a large flush of raw sewage from New York into our waterways as a result of New York City's not replacing its combined sewer and storm-drain pipe system with separate pipes for sewage and storm water. The message is clear: Stay out of New York waters on sunny days that follow a rainstorm, because the water will be polluted when you want to use it most.

Also, once the New Jersey coast was no longer available for sludge dumping, New York sewage sludge was dumped on someone else: shipped by rail to Hudspeth County in West Texas. New York City has promised to build five plants to process its sludge into fertilizer by 1998, something that should have been done a long time ago. The best solution is most efficient and effective when applied at the source.

Connecticut's major sources of pollution were identified as storm-water run-off, sewage treatment plant discharges, and residential septic systems.

Massachusetts has similar primary sources of water pollution: storm-water run-off, sewage treatment plant discharges, and combined storm-water–sewage pipe overflows of raw sewage.

COMPREHENSIVE MONITORING

According to the NRDC report of the 30 states with an ocean or Great Lakes shoreline, only five states had comprehensive monitoring for all their beaches in 1995:

1. New Jersey
2. Indiana
3. Illinois
4. Delaware
5. Connecticut

The nation still has a long way to go to adequately monitor and care for its valuable shoreline.

Lessons Learned (1995)

1. To maintain our current standard of living and our future standard of living we need to improve at a faster rate.
2. We need a quality education, and quality in our education.
3. Taking care of our infrastructure is taking care of our country.
4. A sustainable growth plan requires goals, rules and resources for improvement.
5. We need more than one role model.
6. Continuation of efforts is required to sustain the gain.
7. Similar problems exist throughout the nation.
8. Controlling environmental problems at the source is the most efficient means to reduce pollution.
9. Continuous improvement requires continuous discovery, continuous development, and continuous maintenance.

CHAPTER NINE

Quality Transitions
(1996)

AT&T QUALITY OFFICE—REORGANIZATION

As the AT&T quality office team started 1996, we were focused on completing the transition from the current AT&T to four new companies. The leaders of the new companies would decide whether a quality office was necessary and would determine which people, and resources, they needed from AT&T's current quality office of 26 people.

The split-up of AT&T into separate companies was to help deal with expected increased competitive pressures in a more focused manner. Therefore, total costs for all three companies had to be reduced. Once the decision to establish a quality office in the new companies was made, a zero-based budget approach was used to determine the size of the three new quality offices. Every dollar had to be justified and the total for three new quality offices had to be less than the amount being spent in the current quality office. The three new units already had people working in a quality function in their units who would also compete for jobs in the new quality offices as part of plans to consolidate unit quality staffs with the corporate quality staff. We estimated that the new quality offices would result in only half of our current office staff being offered a job in one of the new quality offices. We advised the entire staff of 26 of this view immediately after the September 20, 1995 spin-off announcement. This allowed sufficient time, the six months from October 1995 through March 1996, for a new job search while we also planned and implemented the transition of the quality office function. By the time we returned from the 1995 end-of-year holidays, 4 of the 26 on our staff had decided to retire.

During the first few months of 1996, the rest of us continued to look for job opportunities, inside and outside the company, while we also supported leaders in the new companies in developing each company's quality office plans. The leaders of the new companies were determining if they needed a quality office, what functions it should perform, how large it should be, and who should be chosen to staff their new quality offices. While our current quality office shared what we did, and how we did it, this was also a time for benchmarking what other companies did and determining what was to be done in the new companies who were making a fresh start.

During this new company planning period, another eight quality office people decided to accept job opportunities in other functions, such as sales or network operations, in AT&T or Lucent Technologies. An additional two decided to accept other job opportunities in other companies. This left us with twelve remaining quality office staff for the potential staffing of the new NCR, Lucent and AT&T quality offices, just less than half the original staff and about the number we had estimated would be required by the new companies.

NCR chose the director, Rob Davis, who had been at the AT&T Universal Card Services unit when that unit had won a Baldrige Award in 1992, as their chief quality officer. Rob added a small staff from existing managers in Dayton, Ohio where NCR had its headquarters. An officer from our network systems group, John Pittman, was chosen as the chief quality officer for Lucent, and a senior officer from AT&T, Frank Ianna, added the role of chief quality officer to his network operations responsibilities.

When the selections of people from the previous quality office were finally made, seven were selected and accepted for assignments in the new AT&T quality office and five were selected and accepted assignments in the new Lucent quality office. Frank Ianna asked me to lead the new AT&T quality office team of seven beginning April 1,

1996, because he already had more than 18,000 people reporting to him for development, deployment, and operations of our long-distance network.

The previous AT&T quality office had its 26 people spread out in four different New Jersey work locations. The use of four locations for 26 people had evolved over the prior few years, due to space limitations in different buildings, as the quality office had grown by adding an office and people in Basking Ridge to support the new chief quality officer, Ken Bertaccini, and an office and people in Parsippany for a new customer satisfaction function. We now had people at the AT&T headquarters building in Basking Ridge, New Jersey along with people to the north in Parsippany, people to the east in Berkeley Heights, and people to the south in Holmdel. Each work location had three to eight people, but the entire new AT&T quality office would have only seven people in total. A seven-person team couldn't work in four different locations and we couldn't afford the rent associated with the space in our current locations. Frank Ianna, my new boss, was located in Bedminster, New Jersey. Therefore, we made a decision to relocate the new AT&T quality office to Bedminster, which meant all seven of us would have a new work location, and most would have a longer commute.

By coincidence, I lived in Bedminster, which gave me a commute of a mile while some of the others on the staff now had to commute 40 or 50 miles each way. While convenient for me, this was a bit embarrassing. Therefore, we decided to set up all seven of us with the same personal computer and electronic mail capability at home that we had at work to allow for telecommuting one day a week, and on bad weather days. This was to offset the long commute people had and to reduce the fear of spending six hours on Route 287 during a snow storm, as some had done during previous years. Route 287 in New Jersey runs north–south through Bedminster, Basking Ridge, and Parsippany, a thread sewing together many of our AT&T locations in New Jersey. Unfortunately, Route 287 had been under construction for three years making our thread a bit knotty. The telecommute agreement for my new office made the transition to the new office easier for the remaining new quality office staff, whom I now desperately needed to stay with me to have a chance to support the quality and customer satisfaction functions we needed to perform in the new AT&T quality office. Cutting to a bare-bones staff meant I now really needed these experienced and skilled people.

Quality Office in a Box

We had also documented the quality office key processes, along with support materials, for use by each new quality office. We prepared a box for each new quality office, with hanging folders labeled and filled with information on each key process. I was insistent that we make this extra effort to provide up-to-date documentation on our key processes for use by the new quality offices. We needed to practice what we preached. Everyone in the office contributed to updating the documentation on their key processes for others to use. Peggy Dellinger, our best writer in the office, undertook the editorial job of ensuring all our documentation was clear, consistent, concise, and complete. Peggy had also been one of the original writers of the first AT&T Quality Library book, *Process Quality Management & Improvement*, and she knew what was needed to adequately document a key process in a usable format.

When people are reorganized, the key process knowledge on how we execute our jobs that has been developed over a long period at great expense, should be documented and not lost in the shuffle. We named the box that contained our AT&T quality office process documentation our "quality office in a box." It contained the experience of the past several years—an intellectual property asset worth millions of dollars. During reorganizations, it is critical not to lose the intellectual property assets of how a company executes its key processes. Losing this knowledge would be equivalent to losing more tangible equipment assets that are also used in executing the key processes of a company. Without either the intellectual knowledge asset or the physical equipment asset for supporting the execution of key processes, a company's results will suffer.

A firm's experience and knowledge on how to execute with excellence is contained in a combination of people, processes, and physical (hardware and software) assets. However, an organization's financial reports recognize only the physical assets. People and process assets are categorized as expenses on financial reports. Without the people and processes (intellectual knowledge), the physical asset by itself cannot produce the required return on investment, and can be reduced to a liability that has to be sold at a loss.

If a manager invests a significant effort to improve the execution capability of an organization by improving the business processes and people skills, the financial report does not show an improvement in the firm's assets. If another manager replaces the first manager, downsizes the organization, and destroys the process capability in a reorganization, the financial report will initially show an improvement. The lower expense as a result of the reduction in people does not reflect the reality that a valuable asset has been lost.

Shortly thereafter, the firm will likely experience difficulty in executing key functions with quality, thereby causing both customer dissatisfaction and expensive rework to correct. The rework must be done by the smaller work force, which in turn causes a drop in morale at the thought of coming to work each day to find you are missing customer objectives, financial objectives, and now have to spend extra time doing rework. When poor results are then reported on the financial report and poor quality execution by individuals is viewed as a cause, the quality approach may be criticized as not working. In fact, if this happens, the quality approach is not working because it was destroyed when the people were downsized and the processes they had built were destroyed in a reorganization that did not maintain the earlier process-execution capability. Reorganizations should be made to support quality execution, which requires careful transition of both valuable people and processes.

When a firm reorganizes and it reduces the number of employees, or transfers employees, it must make an extra effort to maintain key people and the process knowledge. Good process documentation is instrumental in any reorganization. Those of us selected to go to one of the new quality offices knew we would be in assignments where we would have to perform multiple functions with fewer people. With some key people gone, the next best thing would be to have their process documentation. We were helping each member of the current quality office that was selected for an assignment in the new offices and would be expected to do more with less.

REORGANIZATION—SUPPORTING PROCESSES AND SYSTEMS

Reorganizations are usually driven by marketplace changes that require strategy and structure changes, along with budget cuts, as was the case with AT&T in late 1995 and early 1996. Careful transitioning of people and processes is critical to maintaining the execution capability of the reorganized organization.

New market conditions and strategies also require some new product development and some new execution capabilities that did not exist before. These new products, or services, and new execution capabilities require the development of new processes and information systems to support them. A new organization, or reorganization, must focus on creating the accountability, authority, skills, and measures of output to support the new processes, systems, and relationships.

Reorganizations should be designed to support the process-execution capability required by the new market conditions and the new strategy, carefully preserving existing key process elements that are also required for the new strategy. Therefore, processes and systems required by the new market and strategic plan should be designed *before* the new organization design is complete. The new organization structure, or reorganization, should support the new processes and systems required by the market changes and the new strategy, while carefully preserving required execution capability from the past that will also be needed to support execution of the new strategy.

Some companies tend to throw out the baby with the bathwater in major reorganizations aimed at making a major change. The change results in a nice, neat, new organizational announcement and statement of new direction with little execution capability for either the existing business or the new strategy. Unfortunately, many reorganizations are done with objectives like downsizing, consolidating, centralizing, decentralizing, flattening, outsourcing, insourcing, and cost cutting—all of which produce an immediate expense reduction on the financial report. Without organization support for the processes and systems required, execution problems will persist. Execution problems are then frequently blamed on the people who survived the first reorganization and downsizing, quite often leading to subsequent reorganization designs as a solution. Continuous reorganization can be the opposite of continuous improvement, if it is not done with care for both people and processes.

Major reorganizations that involve strategy, structure, people, process, product, service, and system changes, with budget cuts, usually result in a reduced capability to execute the existing business or the new strategy. By applying a quality approach to the transition of existing people and processes and to the development of required new processes and systems, an execution capability can be established to support the new strategy and meet the needs of the new market conditions, while maintaining the required execution capability on the existing business.

Over time the new organizational designs tend to be refined to eventually align with the execution capabilities required to support the new strategy. However, this approach to organizational design, with subsequent organization change and rework, causes major disruption in a company's capability to execute either the existing business or the new strategy, and usually results in performance problems for both owners and customers, and of course for the people involved in the reorganizations.

We can't stop the marketplace from changing and that is what causes the need for strategy and structure changes. However, strategy needs a good execution capability to be successful and structure changes can have a negative impact on execution capability if not done carefully.

A New Approach

It took me a few years in my quality job to realize that what was needed was not for quality to support reorganizations, but for reorganizations to support quality. New organizations and reorganizations need to support the execution of what is needed to be done to support the new strategy through existing and new processes, systems and relationships (internal and external). This approach to reorganization is described in a book that I was asked to review, *Thoughtware: Change the Thinking and the Organization Will Change Itself,* by J. Philip Kirby and David Hughes.

A typical approach to reorganizations is: 1. market, 2. strategy, 3. structure, and then 4. execution. This is an approach supported by many management consultants who focus mostly on the first three—market, strategy, and structure. Consultants tend to do the first three and leave the execution to the employees. Unfortunately, the employees often cannot execute well after a reorganization has destroyed execution capability. Then come the culture-change consultants to change the people, or plan more reorganizations.

The approach needed is: 1. market, 2. strategy, 3. execution requirements (including processes and systems), 4. structure to support execution, and then 5. execution (see Figure 9-1). This is a major breakthrough in thinking about how to address the rate of change that previous quality improvement was not keeping up with. With this approach, we could also do with fewer management consultants. However, we would need more employee

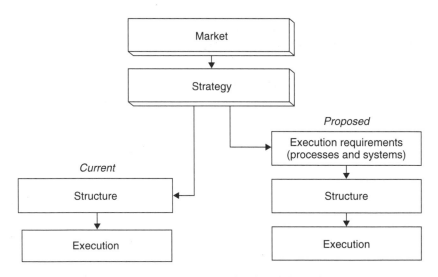

Figure 9-1. New Approach to Reorganizational Planning

involvement in a business that needed continuous improvement of execution.

Toward the end of 1996, AT&T's new president and COO, John Walter, decided to eliminate the overuse of management consultants at AT&T and initiated a process called OWN IT! to support increased employee engagement in running the business. We also increased our focus on the process and system execution requirements for our new strategy.

An old strategy had years to develop good people skills, relationships and processes to support execution. A new strategy needs to develop people skills, relationships and processes to support execution. Change has to be made, but it doesn't come easily and it needs to be managed very carefully. We learned that reorganizations should become a continuous part of the way we manage—to support execution.

We also needed an assessment approach for the capability of our new AT&T business management system (the organization). The Baldrige assessment criteria are the best ones I knew for assessing the effectiveness of an organization's execution capability. However, an assumption in the Baldrige assessment approach is that the management system is remaining relatively constant and is simply on a process-improvement journey over three to five years. To pass the Baldrige criteria, a reorganization needs to support improved execution capability.

A robust quality approach that takes the need for reorganizations into account would have built into it a process for continual reorganization and improvement of the management system to support changing execution needs of the business strategy. A process focus in reorganizations would minimize disruptions of continuous improvement, and improve support of execution.

To support this process focus to continued organizational improvements in the new AT&T, our quality office described the key operational processes and key support processes required to execute the new AT&T strategy, along with the systems and relationships required. Our new AT&T quality office also changed its own quality support processes to support the "new" AT&T.

As a result of the six months of transition planning for the quality office and the time needed by the rest of the company to complete their transition activities, which included a reorganization of the remaining AT&T units and a spin-off of Lucent and NCR later in 1996, it was decided to skip the Chairman's Quality Award assessments of the business management systems in 1996.

The New AT&T

The new AT&T was changing from stand-alone business units for each service (for example, long-distance telephone, high-speed data services, wireless, Internet access, credit card) to a single AT&T market-based strategy and structure for three markets: business, consumer and international. Each market-facing unit needed the capability to deliver a menu of individual communications services, and integrated bundles of services to provide a higher added value to their end customers. For example, a Realtor needs to be easily accessible (when at work, home or on the road) to someone interested in buying a home, and must have quick and easy access to the information those customers need about the area and homes available. Rather than having several, separate AT&T business units selling the Realtors a different AT&T service for part of their communications needs, AT&T's business market-facing organization would develop a menu of services as well as integrated bundles of services that would provide value-added solutions for the Realtors' communications needs. Also the Realtors would receive a volume discount if they chose to receive more than one communications service on one bill. AT&T was reorganizing by combining existing resources that could then deliver a higher value to the customer at a lower cost to AT&T. This was the right thing to do, and we needed to do it right.

AT&T also began to build the capability to offer local service, along with our long distance service, in response to the Telecommunications Act of 1996. This act authorized competition for end-to-end telephone service. End-to-end communications service was provided by AT&T and the Bell System prior to the 1984 breakup of AT&T by the federal government. The 1996 Telecommunications Act was now allowing AT&T to re-enter, and others to enter, the business of providing end-to-end communications service. After developing and deploying our own local service capability, AT&T

could stop paying the high access charges that the monopoly telephone companies charged AT&T on every long distance call.

The new AT&T continued to measure customer satisfaction in 1996 and we set an objective to maintain our level of customer satisfaction performance relative to the competition during the 1996 transition period. Because we had higher customer satisfaction than our competitors did, maintaining this position during a major reorganization would require extra effort on the part of every employee. This objective was tied to executives' compensation to ensure customer satisfaction received the required attention during the period of change. A good part of my time in the second quarter of 1996 involved ensuring that leaders in "new" AT&T organizations understood their responsibility for maintaining customer satisfaction by using a policy deployment approach to connect individual organizations' actions to key attributes that we knew impacted customer satisfaction.

BALDRIGE AWARD CEREMONY

As a result of the November 1995 elections, which put a Republican majority in the House of Representatives, the federal government was also going through its own transition. In a move to force closure on some budget reductions, the Congress shut down the executive branch of government for parts of both the fourth quarter of 1995 and the first quarter of 1996, by not agreeing on a 1996 federal budget. This resulted in federal workers staying home from work, knowing they would be paid later when the budget was authorized. The only people hurt by this tactic were the public that was not served by the federal workers who were forced to stay home. As a result, this hurt the popularity of the Republican Congress' approach to solving the budget problem. It is hard to figure out how forcing workers to stay home and not work, with pay, was going to help solve the problem that the government spends more money than it takes in.

One result of the government shutdown was continued rescheduling of the awards ceremony for the 1995 Malcolm Baldrige Award, when the president would present the 1995 National Quality Awards to the winning companies. The ceremony was rescheduled from a December date to a January date, then rescheduled again to a February date.

The Baldrige Foundation (see Figure 9-2) holds its end-of-year meeting on the morning of the awards ceremony, so that the Baldrige Foundation directors can participate in the recognition ceremony for the winners. Since Bob Allen was the chairman of the Baldrige Foundation in 1995 and I was the chairman of the Baldrige Foundation's support team, we were caught up in the continued rescheduling of the awards ceremony.

When the date was set for the third time, for February 14, 1996, it turned out Bob was not available because of a regularly scheduled meeting with our own AT&T board of directors. Another Baldrige Foundation director agreed to chair the meeting for Bob, with my support. Representatives from the Commerce Department in the National Institute of Standards and Technology informed me that, as the Baldrige Foundation representative from AT&T in Bob Allen's absence, I would be invited to the White House for the awards ceremony. In addition, the president of AT&T's 1994 Baldrige winning unit also was invited, because President Clinton had been out of the country when the 1994 winners received their awards from Vice President Al Gore.

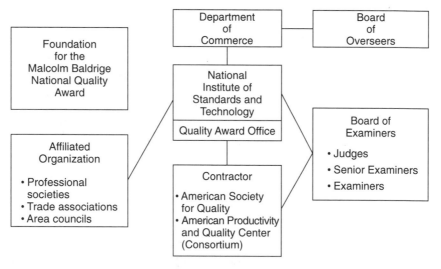

Figure 9-2. Malcolm Baldrige National Quality Award Organization

Red Suit Among Pinstripes

I asked if Judy Soltis, a manager who worked for me and did most of the work to make sure the Baldrige Foundation meeting preparation was done properly, could also attend the White House awards ceremony since she would be flying down to Washington, D.C. anyway to support the Foundation meeting. I was initially told no. However, when the White House staff reviewed who was actually invited to attend, they realized it might be good to have a better representation of women that had been involved in the awards program. At that time, Judy was the only woman directly involved in the Baldrige Foundation or the Baldrige Foundation support teams. A last minute request for Judy to attend was the result.

I gave Judy the good news less than a week before the ceremony and expected an enthusiastic response, which I got. Her second reaction was that she didn't have anything to wear to the White House. That weekend, she went shopping and bought a red St. John knit suit that would be perfect for Valentine's Day at the White House. By the time she bought the shoes and matching accessories, she had spent her entire 1995 bonus check on a once-in-a-lifetime opportunity.

While the red knit suit was being altered, we got another call from Washington. The awards ceremony was being rescheduled again because the president decided to fly out to see flooded areas in Oregon that required federal emergency support. The new date given was in the middle of March. Bob Allen was available that day to chair the Baldrige Foundation meeting and attend the White House awards ceremony. Judy and I knew we would be supporting the Foundation meeting but didn't know if we would still make the White House list, now that Bob was available. The red suit, altered, hung in her closet and it was not returnable.

After several weeks of thinking that the White House visit opportunity had been lost, along with a bonus check on a new suit, we received another telephone call. Judy and I had made the White House list, and the new suit would make it to the White House.

While Judy and I were flying down to Washington with Bob Allen, I told him the suit story. The awards ceremony was held in the Roosevelt room of the White House, filled with pictures of the Roosevelts. While we all awaited the entrance of President Clinton and the start of a formal awards ceremony, the suit story was quickly passed around the room to those who knew Judy. This was a special event being held in a special place with special people. One of those special people was Judy Soltis, a key member of the Baldrige Foundation support team, who stood out in her new red St. John knit suit among a sea of blue pinstripes.

The Quality Approach in Arkansas

In his remarks, President Clinton decided to skip his prepared speech and simply share his personal experience in using a quality approach when he had been governor of Arkansas. He said that a quality approach had been an essential part of efforts to improve the competitiveness of Arkansas industry. He had supported and encouraged sharing of best practices among industry in Arkansas at his annual Quality Arkansas Conference, years before Baldrige started doing the same thing at the national level. He had seen the sharing of best practices pay off in improved economic conditions in Arkansas, just as he had seen the Baldrige effort contribute to improved economic conditions in the nation.

I was probably the only one in the room who had actually met with Governor Clinton at his annual Quality Arkansas conferences and knew that he really had supported the Quality Arkansas effort with his personal involvement as well as support from the governor's staff. When I had the opportunity for a one-on-one handshake

Photo 9-1. Phil Scanlan with President Clinton and Ron Brown in the Oval Office —1996

and photo in the Oval Office after the ceremony, I mentioned my 1989 visit to Arkansas and our sitting together at lunch during the 1989 Quality Arkansas conference. President Clinton seemed particularly pleased to hear that when I returned to New Jersey, I had helped form a similar Quality New Jersey organization. A big smile came across his face and he held onto my hand and reminisced for a few seconds. Perhaps I helped provide a moment of relief in the stressful day of a president (Photo 9-1). When I got home that night, I had a great story to tell about my day in the White House and my second opportunity to meet the president.

BALDRIGE SUPPORT FOR EDUCATION AND HEALTH CARE

The visit to the White House, although delayed, was fun. However, the reason I was in Washington was to support the Baldrige Foundation directors' meeting. My proposal that day was for a transition in the Baldrige Foundation from supporting only business awards to supporting business, education and health care award categories. The purpose was to encourage the same improvement from sharing of best practices that U.S. companies had derived from the Baldrige Award process. While education problems have been discussed, the U.S. health care industry was also facing problems (Table 9-1 and Table 9-2). The proposal, which was supported by the entire support team, included adding four new directors to the Baldrige Foundation board, two to represent the education sector and two to represent the health care sector. The proposal also called for a fund-raising campaign to raise $15 million for an endowment to fund a portion of the expenses associated with the two new award categories. We wanted these two new award categories to also be a joint business–government partnership program with Congress funding a portion of the annual expenses associated with the award. Congress had turned down the proposal for the two new award categories in 1995 when the budget battle was at its height. Therefore, we proposed that the business fund-raising campaign would be to obtain pledges of support, which would be contingent upon congressional approval, and support for the new award categories as a partnership effort.

Table 9-1.

Health Care at a Glance

1990 Figures	Life Expectancy Male/Female	Infant Mortality Rate per 1000	Per Capita Health Expenditures	Health Expenditures as a percent of GDP	Doctors per 1000 people	Average days in-patient care
Japan	75.9/81.8	4.6	$1,171	6.5%	1.6	51.4
Canada	73.0/97.7	7.2	$1,730	9.0%	2.2	13.0
Britain	72.8/78.4	7.9	$ 974	6.2%	1.4	14.8
France	72.7/80.9	7.2	$1,543	8.9%	2.6	12.8
Italy	72.6/79.2	8.5	$1,234	7.7%	1.3	11.7
Germany	72.6/79.0	7.5	$1,487	8.1%	3.0	16.2
United States	72.1/79.0	9.2	$2,566	12.4%	2.3	9.2

Source: "Health Care at a Glance." *Fortune,* July 22, 1992: p. 80. ©1992 Time, Inc. All rights reserved.

Table 9-2.

TQM Health Care Environmental Barriers

1. *A bad outcome* (death) is possible even in the context of good quality care.
2. Lack of incentives to "justify" cost of care or treatment methodologies.
3. The customer has been deprived of the opportunity to assess care.
4. Communications between participants in the health care system are inadequate.
5. Stakeholder exchange is minimal
   - Providers
   - Payers
   - Regulators
   - Accreditation agencies
   - Customers

Source: Reprinted with permission. *Putting the "T" in Health Care TQM.* GOAL/QPC, 13 Branch Street, Methuen, Massachusetts.

Commerce Secretary Ron Brown joined the Foundation meeting that day for a few minutes to thank the members for their support of the business award partnership. He added his personal request for business–government partnership support for the new education and health care award categories that the Commerce Department would be proposing for congressional authorization in 1997, for the 1998 budget.

New Leadership

Based on the proposal of the support team, and Ron Brown's request, the Baldrige Foundation directors approved the plan to raise funds for the proposed education and health care award categories.

The next month, on April 3, 1996, Ron Brown was on a trip to Bosnia with business leaders, including Walter Murphy, an AT&T executive, when his plane hit a mountain in bad weather and all aboard were killed. The purpose of the trip was to help that troubled area make an economic recovery by stimulating economic development with U.S. investments, thereby reducing the potential for continued military conflicts. The request from Ron Brown for support of the new education and health care Baldrige awards was one of his last requests, a request the Baldrige Foundation committed to support.

At the March meeting the chairman of Texas Instruments (TI), Jerry Junkins, was voted to take over the chair of the Baldrige Foundation from Bob Allen. After the March 1996 meeting, I began transferring the support team chair responsibilities to the TI quality vice president, Mike Cooney. However, during the summer of 1996, Jerry Junkins died of a heart attack while on a trip to Europe. The Baldrige Foundation had quickly lost leaders on both the government and business sides of the partnership.

The chairman of Eastman Chemical, Ernie Davenport, agreed to step in and fulfill the remainder of the 1996 Baldrige Foundation chairmanship responsibilities, and continue as chairman through 1997. On the government side, Mickey Kantor was appointed to the post of commerce secretary. Both agreed to support the effort that had been started to establish Baldrige Award categories for the education and health care sectors. I began working on transitioning my Baldrige support team leadership responsibilities to Linda Popwell, the Eastman Chemical quality director.

At the same time, Curt Reimann, the director of the Baldrige Award program in NIST, decided to retire after leading the Baldrige program since 1987. Curt's leadership had been instrumental in attracting and training expert business volunteer quality examiners that provided the rigorous assessments and feedback for improvement to those companies who applied for the national quality award. Above all, Curt demonstrated and instilled a level of integrity in each examiner and in the program that resulted in credibility in the Baldrige Award assessments, the examiners' feedback and the awards.

I had been one of 200 volunteer examiners in 1987 and one of 50 senior examiners in 1988. The three days of training I received each year as a Baldrige examiner were the best that I had ever received. The Baldrige training taught me how to use the Baldrige criteria to assess an entire business or business unit, from leadership to employee involvement, from customer satisfaction to supplier management, from strategic goals to operational results. The training also stressed how to examine for the desired balance in obtaining results for owners, customers, and people in the company, while addressing community issues related to the specific industry.

I not only learned how to examine a unit's capability using the Baldrige criteria, I learned from Curt how to run an assessment and award program, which I replicated in AT&T for our 30 AT&T business units and divisions from 1989 to 1995 as the AT&T Chairman's Quality Award.

Harry Hertz, who had worked for Curt and knew the program well, was selected as Curt's replacement. Harry would continue the Baldrige program with the integrity and credibility that Curt had established.

The Baldrige Award was also in a transition. Global competition was indirectly pushing the rate of this transition. The nation recognized the critical need to improve education and health care as part of the infrastructure support improvement required for U.S. competitiveness. A growing number of us understood how the Baldrige paradigm of management could improve quality and reduce costs in those important sectors, as it had helped the business sector.

AT&T'S QUALITY STRATEGY

The Chairman's Quality Assessment

AT&T converted from 23 product and service business units to 3 market-facing organizations. The new AT&T quality office needed to reassess the quality strategy we had developed and deployed in the old AT&T. For example, running a business-unit Chairman's Quality Award program when we no longer had 23 business units didn't make sense. We decided instead to use the Baldrige criteria to assess our market-facing units' capabilities to execute the new integrated service strategy. However, given the new strategy, structure, processes, systems, and lack of multiple-year results for the new units, this assessment would be focused on driving the necessary development and implementation of improvement plans in key processes, not focused on assessments for awards. An AT&T process framework was developed to identify clear responsibilities and accountabilities for the key AT&T operational and support processes.

To reinforce the new focus on improvement, the Chairman's Quality Award program was changed to the Chairman's Quality Assessment (CQA) program and awards were eliminated from the program. A President's Review step was added to serve as a review by the president of AT&T of the improvement plans that the units would develop and deploy to close gaps found by the CQA assessment process.

We felt AT&T no longer needed an internal quality award to motivate quality improvement. We had learned that a quality business management system produces excellent business results. We simply needed a quality assessment, feedback and improvement program based on the Baldrige criteria.

Customer Satisfaction

Prior to the 1996 trivestiture, AT&T had been measuring customer satisfaction with each service we offered. This supported the business unit organization AT&T had from 1989 to 1995. With a market-facing organization structure it became important to ask a customer how satisfied they were with the entire AT&T relationship as well each service. We realized we now could avoid having the same customer called by multiple AT&T units attempting to understand their satisfaction, thus reducing our survey costs and the potential for annoying a customer with our surveys. We now could get more customer satisfaction information for less money.

Quality Approaches, Methods, and Tools

The AT&T quality office had published 28 AT&T Quality Library books in support of our approach over the past several years. Each provided a how-to approach to applying a particular quality method, including appropriate quality tools. Each book was written for the person or team planning to use a particular quality method or tool (Appendix A).

While we had published this set of how-to reference books in the past, we had relied on each business unit to develop their own approach to quality using these materials and the Baldrige-based CQA assessment. However, we had found that most busy business people did not have time to learn more than a few of the 28 methods we had documented. Some people were using a quality "hammer" to drive both nails and screws, because that was the one quality tool they had the opportunity to learn. Most professional carpenters have a full set of the appropriate tools and know how to use more of them than do weekend amateurs.

The quality office staff realized that we needed to simplify our quality approach because few had time to learn 28 quality methods while on the job. We identified a core quality approach that required only four steps sequenced in a cycle to drive significant improvement in business results:

> Step 1 was a *planning* step, including setting clear long-term goals and annual objectives using benchmarks, and deploying these to all the organizations and people in the company or unit.
>
> Step 2 was an *improvement* or *reengineering* step, for key processes that achieved the objectives and goals in Step one. This step also included careful transition of people and processes when reorganizing, to minimize damage to past improvements.
>
> Step 3 was an *assessment* of the business management system, including key operations and support processes, to measure improvement progress and to get feedback on areas for improvement.
>
> Step 4 was a *correction* step, taking action on the feedback from Step 3 to make further improvements.

A similar four-step PDCA cycle was originally developed by Walter Shewart, an AT&T Bell Labs pioneer in developing quality methods, and used to improve manufacturing process quality. This cycle was called the Shewart cycle and was later used by W. Edwards Deming in his books on quality and referred to as the Deming wheel. Our four-step PDCA cycle extended the PDCA cycle given to us by one of quality's founding fathers, to apply to an entire business management system. By using methods in each step appropriate for improvement of an entire business management system, including Baldrige as a key tool for checking a business management system, we had evolved quality from a set of tools for solving factory floor process problems to a set of methods for solving business management system problems, including all key company business processes.

This core AT&T PDCA cycle approach to improving a business management system uses a few core methods that everyone in AT&T was expected to learn and use appropriately.

- *Planning* included benchmarking and policy deployment.
- *Doing* included process improvement and reengineering, along with a quality approach to reorganization.
- *Checking* included Baldrige criteria for management systems and ISO9000 criteria for processes.
- *Acting* included problem solving.

A 19-page book was written on how these core quality tools were to be used as a system to achieve improved business results. Bob Allen sent the book, via electronic mail, to the top 700 managers in the new AT&T and requested them to read, learn, and use this quality approach in how they managed their part of the new AT&T. The quality office then made the AT&T PDCA book available as part of our AT&T Quality Library as well as on our Internet Web site. The AT&T School of Business developed a training course for the PDCA cycle and the unit quality directors began applying the new PDCA cycle approach to improving their units.

In addition to the core PDCA cycle, a number of useful quality methods in the AT&T Quality Library addressed specific functions, such as supplier management. In 1994 and 1995 AT&T had developed a series of three books that addressed the selection, management, and improvement of a supplier, based on benchmarking the best approaches in many industries. The new AT&T supplier management division applied these quality methods to improve how AT&T manages suppliers. The quality office had also developed a series of three Quality Library books on information management that we expected our information technology systems division to learn and use. The quality office identified the core quality methods that all needed to know and use in a continuous cycle of improvement, and we had identified the functional organizations that needed to apply specific quality methods to improving their particular function.

My Personal Role

Prior to 1996, the business unit quality directors had played key roles in their units in developing and improving independent business unit quality systems. This focus had improved most business units, and helped win several external AT&T business unit quality awards along the way. Many units in AT&T were also recognized for their unit improvement results at our annual AT&T quality conference.

The transition called for by AT&T's new market-facing structure called for integrating many of the capabilities of the previously separate business units and meant a shift to a common quality improvement approach to help the integration and improvement. To support this change in direction, I was given dotted-line management authority to the quality directors in the units as a way to have greater control over the shift from independent unit-specific quality approaches to a common AT&T quality approach. My job changed from being a coach for individual-unit quality managers, focused on improving individual units, to being directly accountable for leading a team of quality managers that needed to work together to improve the entire AT&T quality system.

My personal transition was perhaps the hardest. I had been a coach and advisor to the unit quality leaders, who had been given a lot of freedom in their approach to achieving excellence. Now I had the additional assignment of moving the AT&T quality system in one direction that supported our new business strategy and structure. Each quality manager would have to be motivated to support a common quality approach required by the new market-facing strategy, which required changing what had been successful, but different, unit approaches under the old business-unit strategy. It was time for the business units to give up some of their rights for the good of the company. Because the world we live in had changed, what had worked in the past would not work in the future. The new AT&T couldn't afford to bundle an excellent quality service with one that is not excellent and lose a customer for two services as a result. All of AT&T had to have objectives to reach excellence, all of AT&T had to use a common quality approach, and all of AT&T had to support continuous improvement efforts.

NEW JERSEY STATE GOVERNMENT

In 1994, Governor Christine Whitman approved a state quality program, Quality Innovations, to improve how the state government was run. It received union approval in 1995 and was being deployed in a few state government agencies on a trial basis by 1996. One of the agencies that volunteered to be a trial unit for the new quality program was the Department of Environmental Protection (DEP). The DEP commissioner, Robert Shinn, had seen first-hand the benefits of applying a quality approach to the shore-water quality problems and was willing to try using a quality approach in the other DEP departments that covered land, air, and drinking water quality. An AT&T executive, Nick Finamore, whom the company had loaned to the state for a year, was asked to focus his work on supporting the deployment of the Quality Innovations program to the remaining DEP departments as well as to two DEP-wide processes: the budget process and the permit issue process.

The state's human resources department was responsible for providing the quality materials and training. At their request, I gave permission for the state to reprint the AT&T Quality Library materials for their quality books. They also used the quality department at Rutgers to help conduct their training. Brent Ruben, the quality director at Rutgers and my partner in an AT&T–Rutgers quality partnership, personally conducted some of the state's training classes. Business, education, and government were working together on this quality program.

Nick had a strong background in quality and in managing change. I had known him for years and simply treated him as an additional quality director that I would spend time coaching and advising on how to move his unit, in this case the DEP, along the quality journey. We had the additional advantage with the DEP of having Dave Rosenblatt, who was on my QNJ environment team, as head of the ocean water quality monitoring group in one of the DEP departments. Nick used me, or Dave, to share how we had used a quality approach to clean up the ocean at his kickoff sessions with other DEP department heads. He also found some good work underway in the DEP already aimed at improving key processes, and became a quality coach to support this good work.

Two DEP-wide key processes that were addressed with quality teams were the DEP budget process and the DEP permit process. Dave Rosenblatt was selected to lead the DEP budget process team, due to his years of experience on our QNJ quality team. Nick Finamore provided coaching for the permit process team. Because our QNJ environment team had encountered a number of state budget process problems, I was thrilled that one of our quality team members was selected to lead the work on improving this process. More importantly, the original objective of my QNJ environment team was beginning to be addressed: the application of a quality approach to all environmental issues in New Jersey. A role model example alone is not sufficient to drive a successful approach into other parts of the organization. It has to also be supported by training and management to become an organization-wide approach.

Prior to 1992, the DEP was allowed to keep revenues in their operational budget from fines they levied. Beginning in 1992, fines from the DEP and all agencies went into the state treasury before being distributed by the governor's office. The key improvement in the budget process was to first establish a strategic planning process which established goals and objectives for the agencies. Then the budget process was worked out with the agencies to determine the best allocation of the resources available to enable achievement of those goals. This sounded like the "plan" step in the AT&T PDCA cycle. Establishing goals and allocating resources to achieve them— a great idea for how to run government.

In 1996, for the first time since 1992, the DEP had an approved budget that was higher than the previous year's. This increase was funded by an environment bond issue that passed in 1996, even as Governor Whitman continued cost cutting on a statewide basis. Based on the state's plan for sustainable growth, while protecting and improving our environment, additional resources had to be allocated to deal with a number of environment issues in New Jersey that had been allowed to accumulate over the years (Reference Sustainable Growth plan and NJ environment issues covered in Chapter Eight).

CENTER FOR MARINE CONSERVATION (CMC)

In the middle of all these transitions I was asked by AT&T's environment office to be the after-dinner speaker at the annual International Coastal Cleanup conference run by the Center for Marine Conservation (CMC), held in Washington, D.C. The CMC is a nonprofit environmental group that supports a number of programs to improve the quality of oceans and shores worldwide. One program, the International Coastal Cleanup, is an annual "clean up the beach" day where volunteers around the world pick up all the trash they can find on the shore. However, this program is not about picking up trash, it's about stopping the sources of trash. The organizers ask volunteers to document every piece of trash collected. This data is then analyzed and used to inform people of the types of shore debris being found, and the likely sources.

The CMC then works with the likely sources to make improvements or encourage improvements in any way possible. For example, some cruise ships were illegally dumping trash at sea. The CMC prepared brochures on the problem and gave them to cruise ship passengers asking them to report incidents of trash dumping that they might see while aboard. With thousands of possible "inspectors" on board ships, the incidence of illegal trash dumping declined.

The participants at the annual CMC conference were the state and country coordinators for the annual clean-up day. They came together to share results from the previous year, and plan for the coming year. I spoke about some of my experiences in cleaning up the sources of pollution on the Jersey shore, including our debris. I had also spent an evening analyzing the CMC results reports from the past several years, so that I could address their work. My analysis confirmed that they were doing a great job of analyzing the trash collected and developing action programs to prevent, or reduce, the types of trash found. For example, CMC partnered with the plastics industry and others to improve the biodegradability of materials found in quantity on the beach.

However, I found they were not measuring the overall results of all their programs and they did not appear to have clear measurable goals for debris reduction.

The CMC focus had been on getting volunteers, picking up and collecting data on the debris, and developing and supporting programs to reduce the type of debris found. Their only overall measure of the program was the number of volunteers that turned out each year for the annual clean-up day. Unfortunately, the number of volunteers had not been improving over the past few years. The CMC knew how important this annual event was and were concerned that they were not making the progress on their key goal of increasing the number of volunteers, which disappointed and frustrated coordinators. In fact, the coordinators had spent most of the day before the dinner in planning sessions, working on ways to get more volunteers out for the clean-up day in 1997.

I had analyzed CMC's annual report data for the amount of total trash picked up each year, to see if I could spot an improvement trend, but one did not initially appear. When I trended the trash picked up per beach-mile cleaned, accounting for variation in the number of miles cleaned each year, a clear improvement trend in the amount of total trash did emerge. The audience was thrilled to see this improvement trend that had gone unrecognized. The CMC quality programs were producing overall improved results in debris reduction on the U.S. shores.

The primary goal of reducing the debris on U.S. shores was being achieved, and they had reason to feel proud of their accomplishment and share these results to obtain continued support. My analysis did not show any improvement trend outside the United States, and the number of volunteers and the amount of beach mileage cleaned varied too much in each country to analyze each country independently.

Given this analysis, the goal of the CMC should be for the continued reduction in debris on U.S. shores, instead of getting more volunteers. The number of volunteers was a *process measure*, not a *results measure*. In the United States, we had sufficient volunteers. However, additional volunteers (a process improvement) were required outside the United States to provide an adequate measure of year-over-year improvement and to gain the momentum needed by the people in the country to reduce debris. All the countries in the world together produced only about 60 percent of the number of volunteers that the United States did and cleaned only 20 percent of the beach miles that the United States did.

The 1995 International Coastal Cleanup results report published by the CMC provided data on the size of their annual cleanup effort in 1995.

Volunteers:	U.S.	134,929
	Non-U.S.	79,844
	Total	214,773
Beach Miles:	U.S.	5,870
	Non-U.S.	1,124
	Total	6,994
Pounds of Debris:	U.S.	2,544,009
	Non-U.S.	1,223,247
	Total	3,767,256
Items of Debris:	U.S.	4,057,748
	Non-U.S.	2,143,463
	Total	6,201,211

The Mote Marine Laboratory in Sarasota, Florida is a source of the following information on the biodegradable lifetime of products:

Beverage Containers:	Waxed Milk Carton	3 months
	Tin soup can	50 years
	Styrofoam coffee cup	50 years
	Aluminum soda can	200 years
	Plastic soda bottle	400 years
	Glass bottles and jars	Undetermined
Diapers:	Biodegradable diapers	1 year
	Disposable diapers	450 years
6 pack Beverage Holders:	Photo-Degradable	6 months
	Plastic	400 years

About 60 percent of the debris found in the 1995 CMC cleanup was made of plastic.

The CMC encourages suppliers to use more biodegradable products. Recycling is another important method to reduce debris, along with proper handling of trash and elimination of litter. Litter you see on a street one day will end up on a beach the day after the first rainstorm, via the storm drain system.

Beach Litter Sources

My AT&T quality office team elected to help out on the 1997 Coastal Cleanup Day by picking up litter on the bay side at Sandy Hook. Seven of us picked up 2100 items in three hours along a half-mile stretch of rocky shoreline. That is about one item per foot—which is a lot of trash. The litter included many broken plastic trash bags that seemed to be a likely source of the bulk of the litter. Only a few items appeared to have come from beachgoers' picnics, such as plastic plates and cups. Along this half mile of shore we picked up trash that had seemingly first been put into a garbage bag and then disposed of in the bay, before washing up on the shore. Possible sources of such trash are boaters, people who live on the shore, or the barges that haul trash in the New York–New Jersey harbor. Street litter washed to the bay through the storm drains to the shore is another likely source of the debris we found.

New York City Street Litter

On April 15, 1997, the *New York Times* published their analysis of how well New York City was doing in keeping its streets clean. The *Times'* analysis showed that 64 percent of New York City streets were somewhat littered or dirty. Causes of street litter in New York included such things as:

- Vacant lots and abandoned city-owned buildings which often become dumping grounds.
- Tenants tossing trash out the window into empty lots.
- Property owners not sweeping sidewalks and curbs regularly.

The residents' view was that the sources of the street litter problem in New York City were property owners, the city, and the tenants who litter or who do not regularly clean up their area.

The city Sanitation Department had devoted an intense effort and many resources, including the use of thousands of welfare workers, to clean up the litter on New York City's streets. However, while trash needs to be picked up, clean-up is not a countermeasure for the sources of the problems listed above.

Clean Streets—A Responsibility

Countermeasures are required to motivate and enforce a behavior change by the property owners and tenants in New York, including the city itself. Perhaps a Mayor's Award for neighborhoods with clean streets, like the QNJ beach award we created for New Jersey shore towns, would be one positive motivator. Fines for those who do not regularly keep their area clean is an enforcement of the existing laws that could also be used. Perhaps the fines could be kept in a special fund and used as part of a mayor's clean street award program to fund park improvements for neighborhoods that win the mayor's clean street award. An independent assessment of neighborhoods, using the same standards used by the *New York Times* analysis, could be the criteria for the award.

Quality is all about identifying root causes and developing and implementing countermeasures to prevent, or reduce, the problem. If there are reasons why the city doesn't take care of its abandoned properties, those causes should be addressed. If abandoned properties are not taken care of by the owner, they should be subject to loss of their property rights, with an auction to people who have demonstrated an ability to take care of property. Funds from such auctions could be added to the mayor's clean street award program. Property rights should not be without a responsibility and accountability for care of the property and the environment of the community.

TRANSITION YEAR RECOGNITION—AT&T

At AT&T we were in transition, and therefore had no 1996 quality conference. However, that didn't stop several of our units from getting a little quality recognition for their good work:

1. AT&T Wireless was number one on the J. D. Power Customer Satisfaction Survey of wireless providers.
2. AT&T Universal Card was number one on the J. D. Power Customer Satisfaction Survey for credit card providers.
3. AT&T business communications services received *Data Communications* magazine's User's Choice award for best overall data communications service.
4. AT&T International received *Data Communications* magazine's User's Choice award for the best overall international data communications service.
5. AT&T Language Line won the California State Quality Award for their translation services.
6. AT&T network services division received three International Standards Organization (ISO9000) process certifications on their key processes.

1996 QNJ SHORE QUALITY AWARD WINNERS

QNJ held its annual quality conference and the commerce commissioner, Gil Medina; the DEP chief of staff, Wendy Rayner; and I presented the annual New Jersey Shore Quality Awards. The ocean beach-block-day closings had been held to 10 and the bay's to 75, for a total of 85. This was slightly higher than our best year ever, 1995, when we had a total of 77. The NRDC reported that Florida and California had improved results in 1996, due to reduced hurricane activity.

We had a lot of good work by the shore counties and shore municipalities to thank for our continued good shore water quality results. Our team had decided in our judging that five municipalities and one county deserved recognition for their work in 1996, the highest number of awards ever given. While our QNJ focus group team debated increasing the number of awards, in the end we decided that the award was for meeting our criteria for process improvements that prevented pollution and for achieving excellent shore water-quality results. This award was not about competition between shore municipalities. It was about quality improvements that enabled excellent performance by all our shore towns. We would be happy to have all four shore counties and all 90 municipalities win this award someday.

The 1996 QNJ Shore Quality Award winners included:

1. Avalon
2. Stone Harbor
3. Margate
4. North Wildwood
5. Pine Beach
6. Monmouth County Health Department

The Monmouth County award was for facilitating significant improvements by the municipalities in their county, even though the municipalities were in control of the actual improvements. This change in the award was recommended by Dave Rosenblatt, our DEP member on the QNJ team. Dave realized how a lot of people and groups help support improvement, but in the end it is the municipality that makes the improvements required. This realization is an important point in understanding why, and how, the

New Jersey shore effort had accomplished the most improved shore water quality in the nation. Our effort, and the QNJ Shore Quality Award, focused on supporting, recognizing, and sharing the results and best practices of the 90 shore municipalities, which have the responsibility, accountability, and resources to make the improvements necessary to achieve excellent shore water quality. The 90 shore municipalities were like the 23 AT&T business units that I had supported at AT&T: They achieved the results, the rest supported.

The support staff and support processes are important, but it should be clear which are support and which are the operational processes, and who is accountable for each. The federal government with the EPA, the state with the DEP, the county health departments, and volunteer environmental groups all develop rules, regulations, plans, and measures (with fines and/or awards) and provide expert technical advice (like a corporate staff). But it is the municipalities (like business units) that actually run the operations and produce the results. In New Jersey, we have a strong "home rule" approach to making the final decisions, which makes the shore municipalities the owners of the process and results for our shore water quality. The QNJ focus group team had recognized the shore municipalities as the shore water-quality process owners, and worked to facilitate their support and recognition by those responsible for the support processes. The support processes at the federal, state and county (like a corporation) also have control of resources that can be provided to support the municipal efforts.

The 1996 Baldrige Award ceremony was held on schedule in December 1996. We again held a Baldrige Foundation meeting before the awards ceremony and attended the Baldrige award ceremony. The winners' acceptance remarks were great, and President Clinton remarked that after seeing the success of the Baldrige criteria applied to business, he now believed this was evaluation criteria that should be applied to the way we do work in every sector of the economy, from business to government, from education to health care. While a president's award process already existed to support the application of the Baldrige criteria in federal government agencies, President Clinton was pleased by the proposal from the Department of Commerce and the Baldrige Foundation to jointly support two additional categories of Baldrige awards: health care and education. Given the universally recognized need to make improvements in both the quality and cost of health care and education services, the president encouraged the timely implementation of these new award categories to support improvement efforts in these key sectors of our economy.

The Baldrige Foundation directors and support team appreciated this presidential support for the two new award categories, but we still needed authorization from Congress before we could move forward. In April 1997, the House passed an authorization bill allowing one new award category in 1998, and proposing the second new category be delayed until 1999. Our Baldrige Foundation support team therefore recommended that we proceed with health care followed by an education award, assuming we could also get Senate approval for the House proposal. (See endnote 1 in Chapter Seven.)

Lessons Learned (1996)

1. Ownership for the operations and support processes need to be clearly defined.
2. Root causes and countermeasures are required to prevent, or reduce problems. Clean-up alone isn't sufficient.
3. Support processes need to include both recognition and enforcement for operational processes to help improve results.
4. Reorganizations should be done to improve, not harm, execution capability of processes, people and relationships. The flow should be *strategy*, *processes* for execution of strategy, then *structure*.
5. Our governments should not require subsidies for things we do not want. We should fund things that support our objectives, like support for excellent education and pollution-preventive measures.
6. The municipalities are the process owners for the shore water quality. The counties, state, and federal levels of government need to support and provide leadership for municipal improvement efforts.

CHAPTER TEN

A Culture of Continuous Improvement
(1997)

TAKING A STAND DOWNSTREAM

February 14, 1997, was a Friday. Dave Rosenblatt, our New Jersey coastal monitoring supervisor at the DEP, had thought about taking a vacation day, but he changed his mind and went to work. He was there to receive a call from his counterpart in the New York DEP who informed him that the New Jersey shore might see some pollution over the weekend. New York City was going to shut down a sewage pumping station for repairs. The pumping station was on Thirteenth Street in Manhattan, along the East River. While the sewage pumping station was shut down for four days, 560 million gallons of Manhattan's raw sewage would be dumped, without treatment, directly into the East River. This Manhattan sewage was normally pumped across the East River to a Brooklyn sewage treatment facility. Dave knew that this amount of sewage would create a pollution problem and asked if the plan to dump that much sewage could be canceled or changed. He was told the call was being made only as a "courtesy call." He was being advised of New York's plans and there was no way of canceling or changing those plans.

Having worked on our QNJ environment team for the past eight years, Dave could not just accept as courtesy information intentional plans to dump 560 million gallons of raw sewage in our waterways. This might have been the way things were done in 1987, but this was 1997. We had made a lot of progress over the past 10 years and Dave did not want to see that progress set back. Dave knew that he had to try to prevent New York City's dumping plan that would pollute the Jersey shore that we had worked so long and hard to improve.

Dave immediately checked with the fish and game agency whose computer model analysis had previously showed that 500 million gallons of sewage released all at once in the New York–New Jersey harbor would pollute northern New Jersey shellfish beds. Next, Dave notified the New Jersey DEP commissioner, Bob Shinn, of New York's plan and of the shellfish pollution problem that would result. Commissioner Shinn immediately issued an order to close New Jersey's shellfish beds to fishing as a precautionary move, and he called New Jersey's governor, Christine Whitman.

Governor Whitman did not appreciate the 560 million gallon Valentine's Day present that New York had planned for New Jersey. Following her angry, Friday evening call, New York's governor, George Pataki, agreed to postpone the dumping, which had been set to begin on Saturday, February 15, 1997.

The plan had been postponed by Governor Pataki, but not yet canceled or changed. Governor Whitman held a press conference that Friday evening to inform the public of New York City's plan to pollute the Jersey shore. Public awareness and concern for this dumping would help ensure that New York City changed its plans. Governor Whitman told the press, "This, obviously, is totally unacceptable. It's mind-boggling that anyone would consider it. It is utterly medieval when you talk about putting this kind of raw sewage into the waterways." Commissioner Shinn then added, "We have never seen a quantity approaching this level. This is unprecedented." The last time that closing such a large pumping station required raw waste to be dumped was in 1987 when 200 million gallons of raw sewage were dumped, according to New York City officials. Apparently in 10 years Manhattan had more than doubled its capability to produce raw sewage, but had not adequately improved its capability to handle it.

Mayor Rudolph Giuliani, who was interrupted while watching a Friday night movie, was quoted in Saturday's *New York Times* as saying that he did not believe there was anything wrong with the dumping plan, although he had not been notified about these plans prior to getting a call from the governor. Giuliani emphasized that as far as he knew, New York City's Department of Environmental Protection had abided by all state and federal laws in planning the dumping. "It appears the city is doing precisely what it is entitled to do, and New York State DEP had approved it," Mayor Giuliani said.

The mayor was partly right. The federal regional EPA office had given the New York State DEP authority to "permit" such dumping of raw sewage into waterways and New York State had, in fact, given the city "permission" for this particular dumping plan. Therefore, the city was acting with the permission of both the state and federal levels of government. However, the potential impact on the downstream state of New Jersey had not been considered in this planning.

I had been at Tufts University the previous weekend working on this book. While using the library a Tufts student, Andrea Lubin, who was aware of my book from conversations with a Tufts trustee, suggested I review the books she was reading for her Tufts course on environmental policy. During a break from writing I visited the campus book store and purchased the required set of five books for the environmental policy course. One was *Controversial Issues in Environmental Policy* by Kent E. Portney, who also happened to be the course professor. During lunch, I read the chapters on water quality policy and discovered an interesting provision in the federal Clean Water Act amendments of 1972. One provision enables the EPA to revoke a state's right to issue permits for pollution discharge if it violates EPA guidelines or if a governor of one state reports that his or her state will be adversely affected by pollution from another state. I was surprised to find this provision and thought it might someday be useful in establishing New Jersey's downstream rights if New York continued to be a source of pollution on the Jersey shore. I didn't think that would be the following week.

After the weekend press conference and the postponement of the dumping plan, a meeting was scheduled for the following Wednesday between New York, New Jersey, and regional EPA environment officials to discuss the plan. After reading about the issue, I sent Dave Rosenblatt, New Jersey DEP commissioner Bob Shinn, and Peter Brandt (our QNJ team member from the EPA New York–New Jersey regional office) the information on the rights of Governor Whitman to appeal to the EPA on this issue using the Clean Water Act provision of 1972 that was identified in Professor Portney's book. This provided New Jersey with a legal option if logic and public pressure did not get New York to agree to change their plans.

I also provided a few suggestions for possible ways that New York could reduce or eliminate the planned raw sewage discharge, based on similar techniques AT&T would use to repair one of our AT&T switching centers without "dumping" the communications traffic we carry on our network. Our glass fiber cables carry communications; sewer pipes carry sewage. Our switching centers direct the calls to the right place; pumping centers move sewage to the right place. Since a pumping station performed a function similar to our switching centers, our procedures for repair at a switching center could be used as a model:

- Install a back-up pumping capability to enable pumping to be switched to the back-up pump when routine repairs are required, or in the case of a pump failure.
- Schedule the repair work only in the early morning hours when the flow is light. Store that light flow in a river barge for pumping after the repairs have been completed. AT&T only stores and forwards certain types of data traffic that can stand a delay in transmission. Based on the customer importance of the reliability of delivery compared to the importance of speed of delivery, sewage could stand a bit of delay.

The initial reaction of the New York officials to these suggestions by Dave was to indicate that they cost money and that money for this was not in their budget. Dumping the raw sewage into our waterways had a zero cost to New York. The cost was New Jersey's, from pollution of the downstream shellfish beds and our beaches.

Due to Governor Whitman's pressure and the attention she drew to New York City's plans, New York reluctantly agreed to postpone the dumping until at least December 1997. This would allow time to study possible alternative solutions to the repair problem and allow both the EPA regional office and the New Jersey DEP to review the revised plans. An unenthusiastic Mayor Giuliani was quoted in the February 19, 1997 *New York Times* as saying, "Out of an excess of caution, there are probably additional review processes that we can go through."

Mr. Sheffer, a spokesman for the New York DEP, said, "We did not send a written notice to New Jersey or to the Interstate Sanitation Commission, which coordinates sewage policy in New Jersey, New York, and Connecticut. We were wrong, it's clear that our notification procedure needs some improvement." Joel A. Miele, Sr., New York City's environmental protection commissioner, said he had not realized that the sewage dumping was imminent. On February 19, 1997, the *New York Times* quoted Commissioner Miele as saying, "I didn't know the planned discharge was about to come up." The culture in New York was such that the dumping of 560 million gallons of raw sewage into our waterways was not considered significant enough to bother informing New York City and state leaders.

In 1997 the New Jersey ocean water quality was no longer a one-party political issue. It was a "Jersey" issue. The Jersey shore is what Jersey is proud of and what Jersey people love most about New Jersey. New Yorkers have their fantastic city with great plays, museums, operas, and other forms of entertainment. New York is an exciting city to visit, it is New York's tourism attraction, and New Yorkers love their city the same way Jersey people love their shore. The shore is New Jersey's tourism attraction. New Jersey would no longer let New York, or anyone else, dump on our shore. We had developed a culture of continuous improvement at our Jersey shore, and we would stand up to anyone that tried to set us back. Our knowledge of our downstream rights under the Clean Water Act of 1972 helped when all else failed. Those downstream had the right to take a stand and knew it was the right thing to do.

During this same period, New York City, under the leadership of Mayor Giuliani, had made significant improvements, including a major reduction in the crime rate. New York was improving the city they cared about while New Jersey was improving the shore we cared about. However, New Jersey was downstream from New York and needed help from New York with our efforts to prevent shore pollution.

Upstream and Downstream Cooperation

In business some functions are "downstream" from other functions and also need downstream rights of appeal for correction of poor upstream quality. For example, at AT&T we had begun to use a simple model of our customer experience as a way of integrating our different service capabilities in the new AT&T, which we called LB GUPS: Learn, Buy, Get, Use, Pay, and Support. Each one of these experiences required an internal AT&T process to support it and each process was sequentially downstream from the one before it. If the quality of the order input in the Buy process is not perfect, it has no impact on the salesperson's commissions but it has a major impact on the rework and cost that results in the Get (provision), Use (communications service), Pay (bill), and Support (customer service) processes of the company and in the quality of service that the customer gets. The downstream processes need the right to appeal and require corrective action if the sales function does not have the highest possible quality input. From the salesperson's perspective, it may appear their commissions are highest if they spend little time and costs on the quality of the order input and all their time on selling new orders. Order input time may be viewed as administrative, nonproductive time. Dumping poor quality orders into the process is a zero cost impact to the upstream sales function, in the short run, like dumping raw sewage into the river. However, if repeat customer orders are negatively affected by poor quality service, repeat customer sales will be reduced. That is a longer term effect and if the average time for a person in a particular sales job is less than two years, the longer time may never happen during the tenure of the individual salesperson, but the company and the salesperson's replacement feel the impact. The sales function situation is similar to that of New York (upstream), and the customer service function is similar to that of New Jersey (downstream). Each is managing its own piece of the stream. However, pollution flows downstream and the downstream people need to have some rights to stop the upstream pollution. A potential incentive to improve sales order input quality would be to reduce the sales commissions by a percent that is equal to the percent of orders that are input with an error. For example, if 20 percent of the orders were input with an error, the salesperson's commission would be reduced by 20 percent. Upstream behavior must be motivated to reduce the negative impact on downstream functions.

Over the past few years we had tried to partner with the New York DEP by inviting them to our QNJ team meetings, which we even held a few times in New York, or on the New Jersey side of the Hudson River at Liberty State Park. However, we had not yet been able to get New York participation. This is also the case in many business situations when a downstream function is dealing with an upstream function. The time it takes to partner and solve problems will not happen until all parties recognize they have a common goal, which is to keep the whole stream clean.

Before our next meeting, I asked Dave to again invite his counterpart in the New York City DEP. I was hopeful that this time someone would show up. Initially we were told that someone would attend but, at the last minute, concerns were raised about working with an outside group, and again no New York City representative participated in our planning meeting.

We couldn't work on possible solutions for the pumping station repair problem without New York's involvement. Therefore, we decided to analyze the plan for the replacement of combined sewer overflows (CSOs) on the New Jersey side of the river, which also provided a direct overflow of raw sewage into the river, whenever we had heavy rains. Replacement of these CSOs would be an expensive project costing billions, but if it were spread out over the next 20 years it could be done—and it needed to be done. If we had started in 1972, when the Clean Water Act was passed, it would be done today. Dave agreed to request review of a DEP plan to repair CSOs that would cost half the replacement amount but only result in a 10 percent reduction in pollution. Replacement of CSOs looked to us like a much better investment than repair at this point, even if it took longer. In addition, we wanted to take away an excuse New York had used to continue their dumping of raw sewage: New Jersey still had CSOs that resulted in raw sewage overflow.

A PARTNERSHIP FOR COASTAL MANAGEMENT

While we had not yet enlisted New York as a partner in our improvement efforts, our New Jersey governor was trying to improve the state's partnership with the shore counties, the shore municipalities, and the shore environmental associations. On April 29, 1997, the governor held a Coastal Alliance Conference in one of our shore towns, Long Branch, to review a New Jersey DEP report, called "A Framework Document for a Coastal Management Partnership," with the shore counties, municipalities, and environment associations.

Two to three hundred people attended this conference whose agenda included the DEP's presentation on their report for coastal management partnerships. Small group breakout sessions allowed opportunities for reaction and advice. Both Governor Whitman and DEP Commissioner Shinn attended each of the discussion sessions. This partnership document and the personal participation of the governor and the DEP commissioner in listening to comments from those at the conference was another significant sign of the new culture of support for continuous improvement of the New Jersey shore.

The 125-page report was an excellent framework document containing numerous facts on the shore water quality, including a description of many current issues and plans to address those issues. In addition, the DEP provided a CD-ROM digitized map and database of environmental information that every participant could take home and load on his or her computer for further analysis of a particular shore area, or town, of interest to them.

The purpose of the framework document and the conference was to provide the current facts, data, issues, and plans as a basis for developing a "New Jersey Coastal Partnership Management Plan" to continuously improve our shore, both economically and environmentally, with the support of all stakeholders. The stakeholders included federal and state governments, shore counties, shore watershed regions, shore municipalities, shore environment associations, and citizens who lived on the Jersey shore.

The framework report acknowledged the following six key barriers to continuous improvement that needed to be addressed.

1. Inconsistent Regulations at Various Levels of Government

The perceived problem was that various levels of government, as well as other public and private entities, traditionally have operated under separate mandates, and with different priorities and diverse agendas, particularly where coastal issues were concerned. The regulated public often expresses concern and frustration about the overlapping (and sometimes contradictory) rules, regulations, and policies between federal, state, county, and municipal governments, and the use of standards that are inconsistent among agencies.

The report indicated that while significant progress had been achieved, incremental loss and degradation of coastal resources nevertheless continue. The report included a deep concern about the capacity of the coastal environment to function as a healthy, productive ecosystem. That capacity was reported to be declining as a result of increasing population growth and population densities. The shore issues were being addressed by multiple levels of government which were generating multiple plans, which included a federal plan, a state plan, a coastal area plan, four shore county plans, nine regional watershed plans, and ninety shore municipality plans.

While support for the development of plans was strong at the higher levels of government, New Jersey, like many other states, leaves most of the authority, accountability, and responsibility for implementing plans to the municipal level, as we do with our education system. Anyone can have a plan, but the municipality has the "home rule." We needed the 90 municipalities to be aligned to make the changes required for continuous improvement of the Jersey shore. Many levels could develop plans, but it was the municipalities that needed to deploy these plans. Most other levels measured, studied, planned, advised, financed, regulated, and supported the municipalities, but the municipalities were the implementers. The municipalities were the "business units" and the other layers of government were the "corporate staffs." Understanding the importance of the municipality, our primary level of government, is a key to making the improvements required. Many expect the federal or state level of government to be the key to making improvements, but they are not. It is at the municipal level that we make improvements. The federal, state, and county levels of government need to support and encourage the municipal improvement effort.

One of the proposals for protecting the shore environment, while supporting continued population and economic growth, was to channel the growth to areas that already have infrastructure support. This would provide the most cost effective growth for the municipality and also protect the shore's limited natural resources. While this is a simple and logical idea, it goes against the idea of American individual freedom to do anything we want with "our" land and "our" individual property rights. The required common rights of the community to protect the rapidly reducing common natural resources was not even a consideration when the Constitution was written more than 200 years ago. The community's common right to protect the diminishing natural resources of this country is an issue that now has to be dealt with. Some proposals included the right for the community, at the municipal level with state financial support, to purchase the shore natural resources that we wanted to save as a community, a coastal "Blue Acres" program to supplement the "Green Acres" program that had been developed in prior years to save farmland. These purchases would be from voluntary sales, but they also could include the swap of government-owned land away from the shore for shore property as an alternative to purchase.

The state also planned creation of a "Brown Acres" program to purchase abandoned lots and buildings in the urban areas to restore the inner cities and to later resell these "Brown Acres" for development, which would produce jobs in the inner cities. Green, blue, and brown acre programs, without a lot of red tape, could do a lot to save our farmlands, our shores, and our inner cities. The community has the capability to save the community if they want to pay for it. These green, blue, and brown acre programs required financing, which were supported as bond issues by New Jersey voters in 1995. New Jersey's citizens decided to invest in their state, to help save and restore what was important to them. The 1995 investments were a start, but ongoing investments would be required in the future.

2. Sea-Level Rise

Sea-level rise is another significant long-term shore issue. During the last century the sea levels along the Atlantic coast have risen 12 inches and are projected to rise more rapidly in the next century. When the seas rise, homes become more vulnerable to storm and tide damage and recreational beaches are diminished. In addition, the ends of the storm-drain pipes may end up under water, thus causing an inability to drain during a storm, which can then cause flooding at the inputs.

Mitigation techniques to protect the shore property as the sea level rises are important, but expensive. They include such things as beach replenishment, barrier walls, land use management (for example, zoning or setbacks) and "Blue Acres" acquisitions.

Repair of beaches, building barrier walls to separate us from the ocean we want to see, and rebuilding storm-drain systems is a very expensive way to fight sea level rise. Mitigation is a repair strategy, which is necessary and required by the Federal Emergency Management Agency (FEMA) as a condition for receiving federal disaster assistance. Mitigation is required by the Robert Stafford Disaster Relief and Emergency Assistance Act. Whereas pumping sand onto the Jersey beaches initially looked to me like an expensive waste of money, it was required by the federal government and partially funded by the federal government to provide some protection for shore property and to qualify New Jersey for disaster-relief money, much like a requirement in an insurance policy to have smoke detectors in a home to minimize damage from a fire.

A quality approach always looks for a preventive alternative to repair work, which is usually less expensive. In this case, we needed to prevent, or reduce, the rise in sea level, to avoid all the associated mitigation and repair work. While this sounds like an impossible job, so did cleaning up the Atlantic Ocean along the Jersey shore when we started a decade ago. Sea-level rise is a very long-term problem, which gives us time to deal with it, but we have to start. If we could reduce the root causes for the sea-level rise over the next decade, we could avoid, or reduce, the associated mitigation problems and expenses that would result if we did nothing to prevent the sea level rise.

The cause for the sea level rising is known. It is caused by the rising temperature of the earth. The average temperature of the earth has increased by about five degrees centigrade since the last ice age, over 18,000 years ago, when glaciers covered much of the surface of the earth, including parts of New Jersey. This five degree change has resulted in a 300-foot increase in the sea level over the past 18,000 years, 1 foot of which has occurred in the past 100 years. Higher temperatures result in the melting of glaciers, which causes an increase in ocean water volume. In addition, the higher temperatures cause the sea water to warm and expand. While many glaciers have melted, the glaciers in Greenland and Antarctica still contain enough water to raise global sea levels by an additional 200 feet, which would certainly give us a lot of very blue acres, without our spending any money to purchase them.

Instead of preventing the cause of sea-level rise, it turns out we have been increasing it. The global temperature has been increasing at a faster rate over the past century principally due to the increase of manufactured gases that prevent the venting of heat into space. The most significant of these "greenhouse gases" are carbon dioxide, a product of fuel combustion, and methane, a product of landfills. Carbon dioxide accounts for approximately half of the global warming that has been experienced to date and for half of the projected global warming.

It would be nice if we knew how to repair our atmosphere, but we don't, and greenhouse gases once released stay in our atmosphere for 100 to 1000 years. Perhaps this atmospheric repair work would be a good challenge for our space program. When President Kennedy challenged our space program to put a man on the moon we did not know how to do it, but we did it in nine years.

Spaceship Depreciation

Putting this greenhouse gas into space is causing damage in our "Spaceship Earth" that future generations will inherit—a destruction of the earth's assets that go unaccounted for in our GDP calculations. There are no write-offs or depreciation expenses in the accounting methods used by our various governments when they report how well our spaceship is doing. The depreciation and write-offs to recognize the destruction of valuable spaceship assets go unreported. A business CEO could not get away with such reporting because the "generally accepted accounting principles" require reporting if current results are obtained by using up valuable business assets. Normally the current results reported are reduced by subtracting the amount of assets used, through write-offs or depreciation. Government reporting should also include an amount for depreciation of the assets they are responsible for taking care of. We should account for the use of our natural assets when reporting current results—it is a real cost that presently is unaccounted for. In addition, some of our natural assets are not as easily replaced as business inventories, equipment, and building assets. Government assets should include our land, air (including outer space), and water quality. If they are degraded, their value is reduced. If their quality is improved, their value is increased. Governments need a means of accounting for the balance-sheet value of the environment, and increases or decreases in environmental assets for which they are responsible. The dumping of millions of gallons of sewage into our waterways and the pumping of greenhouse gases into our atmosphere have a cost in terms of depreciated environmental assets. This cost needs to show up on the "books" of today's governmental managers, not just on the doorstep of future generations.

At the 1996 Conference of the Parties to the Convention on Climate Change in Geneva, Switzerland, more than sixty insurance companies signed a statement asking governments to substantially reduce their emissions of greenhouse gases. They provided data that showed that worldwide economic losses from weather-related natural disasters had risen to $38 billion in 1995 from $5 billion in 1985, almost 4 times the rate of inflation—the beginning of accounting for the cost associated with greenhouse gases that have polluted our environment.

Reducing Emissions

The 1992 International Framework Convention on Climate Change in Rio de Janeiro, Brazil produced a protocol urging countries to reduce their greenhouse gas emissions. Since then, the United States and more than 160 other countries have ratified an International Framework on Climate Change, which establishes a goal for developed countries to reduce greenhouse gas emissions to 1990 levels by the year 2000. Unfortunately, the goal set for the year 2000 will not be achieved. Therefore, President Clinton has committed the United States to develop a policy position for a 1997 meeting in Kyoto, Japan, where new goals will be set for the greenhouse gases for the years 2010–2015. This policy position will require verifiable and achievable reductions in greenhouse gases in a flexible and cost-effective manner. Setting goals is an important step but the pollution process must be managed and preventive steps taken to reduce the sources of pollution. This process also needs a process owner and a process improvement team.

While scientists do not agree on the forecast for the sea-level rise over the next century, the U.S. Fish and Wildlife Service and the University of Maryland have collaborated on a report, "Vanishing Lands," that examines the effects of global warming and sea-level rise on Chesapeake Bay. Recommended options for addressing the problem of sea-level rise in the bay included coastal defenses that are built to withstand a potential two to three-foot rise in sea level over the next century. Repairing beaches and building barriers is expensive and does not address the root cause of the problem. We need to reduce our consumption of fuels that produce carbon dioxide and ozone pollutants by reducing our overall fuel consumption and by changing to fuels with less pollutants. Perhaps to help in the understanding of this problem we should also change the name greenhouse gas to bluehouse gas, since a greenhouse produces green plants and these pollutant gases will produce a blue, not a green, planet.

To simply reflect today's costs associated with increasing shoreline damage due to sea-level rise, we should substantially increase the tax on fossil fuels (by at least a dollar a gallon over four or five years) to encourage the use of alternative fuels. The additional tax money should be directed for exclusive use by FEMA emergency relief, and for shore mitigation efforts, including blue acre purchases.

New Jersey is one of 30 states that has completed work on greenhouse gas emission inventories. Twenty states, including New

Jersey, have also begun work on action plans for their amelioration. Over 80 percent of New Jersey's greenhouse gases are from carbon dioxide, which is similar to the levels that other states report on their inventory. Transportation is a source of 38 percent of New Jersey's carbon dioxide, while fuel consumption in residential homes accounts for 23 percent. Electric power generation accounts for 22 percent and business fuel and power generation account for 16 percent.

Both carbon dioxide and ozone are pollutants released when fuel is burned to power an electric generator or power a vehicle. Ozone is known to cause respiratory problems and is a major cause of air quality problems. If we reduce our fossil fuel consumption we will improve our air quality and save our shore.

3. Algae Blooms

The New Jersey Coastal Report also contained a report on algae bloom levels in New Jersey coastal waters, indicating that intense red tides, like those in North Carolina, occur several times each summer in the Shrewsbury River, which feeds into the northern part of the Jersey shore. These red tides are occasionally severe enough to also result in fish kills. While the source of the fertilizer that produced the red tide in North Carolina was hog farms, the Monmouth County race track and horse farms were the likely sources in New Jersey. Therefore, the Monmouth County health office was working with both to make improvements. The more we pollute an environment, the more the life in that environment responds by becoming tougher.

4. Boater Sewage

Boater sewage was another source of pollution addressed in the coastal report. In 1990 the United States Coast Guard reported 152,000 registered recreational boats in New Jersey. If just one-third of these boats (50,000) discharged their daily volume of raw sewage (20 gallons) into our waterways, it would equate to one million gallons per day of raw sewage pollution (twice the amount of the planned New York City dumping) by the very people who wished to use these waters for recreation. New Jersey, therefore, increased its number of mariner waste-discharge stations and increased awareness programs to encourage the participation by all boaters in the program for proper discharge of sewage.

5. Storm Water Runoff

Another key area covered in the report was proper planning to reduce pollutants from storm-water runoff. In New Jersey, 20 watershed management areas were created based on an analysis of the development and topography of New Jersey. Five of these watershed management areas were along the Jersey shore. Proper plans developed by them could reduce the runoff and reduce pollutants from entering the 150 municipal storm-water discharges to the ocean and the 7000 municipal storm-water discharges to the bays. While we organize our government structure to serve communities of interest, the watershed areas are defined based on how best to manage the flow of water, which does not stop at a town's border. These watershed plans for how to best deal with water flow can then be shared with the counties and towns in the watershed area. We expect municipalities to make and enforce plans and zoning that support the best results in the watershed. Again, we have another level of planning which requires the municipality to align and deploy its plans in the best interests of the broader community. While logic is useful here, it also requires both positive and negative incentives to get all to align with the common good as a priority over the individual's or municipality's best interests. If municipalities are competing for development and jobs, then the concern for the common best interests of the environment could be placed as a distant second priority.

6. Independent Individual and Municipal Decisions

All these issues, and plans to address them, face a similar barrier. How do we get individuals and municipalities to put long-term common-good interests ahead of their own best short-term interests? The DEP report did not address this barrier, but did call for a partnership approach to address it.

Unfortunately, I have found that partnership efforts never occur unless the common goal is made clear and made a requirement, not an option. Protecting our environment can no longer be optional. It is the job of Congress to set goals for improvement in our ocean water quality that we all should work to meet. How we do it should be based on input from all and in a partnership mode. It cannot continue to be an independent decision of each individual and municipality. Individual property rights and municipality home-rule rights have to be put in balance with the right of the community to protect the earth's environment, which is being degraded each year, impacting the rights of future generations—future generations that have no vote in the actions of today's generation.

As with our budgets, we should be at least in balance on our key environmental indicators. We once thought of our environmental assets as limitless and free. Most now know that is not true, but some like to still act that way because there is no cost today to dump sewage in our waterways and no cost today to continue to pollute our air and space with greenhouse gases. Today's costs for using the environment need to be raised to reflect the real environmental asset loss, as we would do in generally accepted accounting practices for business when we use up an asset to produce today's profit. If we established a per-gallon charge for dumping sewage into our waterways, a budget could be established for replacing the antiquated parts of our sewage and storm-drain systems and we would discourage the motivation to dump raw sewage, because it would be cheaper to build an adequate sewage and storm drain network.

NEW JERSEY SUSTAINABILITY

On May 8, 1997, a nonprofit organization, New Jersey Future, held a follow-up Sustainability State Leadership Conference on New Jersey Sustainability Measures and Means, at Princeton University. Governor Whitman was again the keynote speaker, as she had been at the initial Sustainability Conference held in 1995 at Princeton. As in 1995, I also was invited to attend this conference to represent Quality New Jersey.

In 1987 the Brundtland Commission (the World Commission on Environment and Development) defined sustainable development as development that will meet the needs of the present without compromising the ability of future generations to meet their needs. The purpose of this New Jersey conference was to propose, and get comments on, a set of indicators of sustainability for the state of New Jersey. These sustainability indicators would measure whether New Jersey was making progress toward a goal of achieving sustainable development.

If we had a balance sheet for planet Earth, sustainable development would equate to maintaining our valuable environmental assets. We would not have a net depreciation and write-offs (degrading or using up) of our environmental assets each year.

The twenty sustainability indicators proposed for New Jersey covered more than just the environment and addressed the economy and our society as well. The proposed environmental indicators of sustainable development were as follows:

1. Shore water quality: Ocean and bay beach-block-days of closings.
2. Drinking water quality: Percentage of water systems contaminated.
3. Air quality: Trends in air pollutants.
4. Traffic: Total annual miles traveled per mile of road.
5. Energy consumption: Btu per capita per year.
6. Ecosystem health: Breeding bird populations.
7. Respiratory disease: Asthma cases per 100,000 residents.
8. Waste generation: Tons per capita per year.
9. Farmland: Acres of farmland.
10. Quality of life: Percentage of residents rating their community as a good place to live.

The first three—shore water, drinking water, and air quality—had been improving over the past several years. The shore water was the most improved, by a factor of 10, over the past decade. The drinking water had improved by a factor of three over the past decade and air quality had also improved somewhat. These areas could be thought of as an increase on the balance sheet of environmental assets for New Jersey. Investments in these improvements made our home in New Jersey a nicer place to live and made our New Jersey home worth more for both current and future generations.

The next two—traffic, and energy consumption—had shown no improvement. Improvement in the level of traffic and energy consumption was needed for sustainable development. We had not made the needed improvements in our New Jersey home in these important areas. Traffic and energy consumption were problem areas that still needed to be addressed and improved, as had been done with our ocean water quality problem.

The next five New Jersey environmental indicators were all moving in the wrong direction. We had increasing waste per person, increasing rates of respiratory disease, and decreasing bird populations. The decreasing bird populations, besides being a concern to those who like birds, are a warning signal, like the canary that is used in the coal mines to signal danger from coal gas when it dies. We also had decreasing farmland acreage as a percentage of New Jersey's total acreage.

Finally, the percentage of New Jersey residents who felt their community was a good place to live declined from 72 percent in 1988 to 70 percent in 1995. This last measure, which is a perception by New Jersey citizens, is an important measure of the perception of the customers of the state, and a score of 70 percent is a low C.

Based on customer satisfaction work at AT&T, when we do not get an excellent rating on our customer satisfaction survey we have a high loss of those customers to competition. AT&T customers have come to expect excellence from the AT&T brand, and when we disappoint them by providing only good service they try a competitor to see if they can get good at a cheaper price. That is why we have set our goal at AT&T for excellent customer satisfaction: Good is not good enough to hold AT&T customers in a very competitive environment.

New Jersey also needs to establish a goal for excellence in the New Jersey residents' perception of the quality of life in the state, and set annual objectives for improvement from our score of 70 percent. This C− needs to be raised to a C, C+, B−, B, B+, and then to an A over the next several years.

It might be tempting to say that we almost had a sustainable situation in New Jersey, with three positive indicators, two flat and three negative environmental indicators. Unfortunately, however, we need clean land, water, and air for sustainable development. We need all ten to be improving for sustainable development in New Jersey and we had only three of ten improving.

The conference also proposed another ten measures as indicators of the economic and social well-being of New Jersey, such as the growth in the Gross State Product (GSP). This is a state measure similar to the national Gross Domestic Product (GDP). While this measure was improving, indicating economic improvement, it does not contain an expense for the depreciation cost of the environment that is indicated by the results achieved on our environmental indicators. Thus we are leaving to future generations the cost of cleanup and the cost of putting in a system to protect the environment. (This is one of the costs that Japan is now paying by building a sewage infrastructure that they had put off building in the 1960s and 1970s when they were only focused on economic growth. This delayed infrastructure investment cost is one of the major reasons that Japan ran a large economic deficit in the 1990s.)

Improving the Sustainability Indicators

Our success with improving the shore water quality gave me confidence that applying a similar quality approach by dedicated citizens, in partnership with government, could produce similar improvement results on all the key sustainability indicators. The potential for broader application by teams was also made possible by building on the state's Quality Innovation (QI) program. With the help of more citizen-led focus group efforts, using the Quality New Jersey approach to supporting improvement in government, partnerships could change that low C to a B or an A on New Jersey's sustainability report card. Meaningful civic engagement is required to promote excellence in government. With involved citizens who care about improving results, New Jersey could set and achieve goals for excellence as had been done with the shore water quality. If New Jersey had similar joint government–civic partnerships on all 10 of the New Jersey sustainability indicators, goals for excellence on each, and teams committed to achieve excellence, they could do just that.

Use of less fuel, more efficient fuel-consumption vehicles, or alternative fuels could also pay off for the environment, in reduced greenhouse gases, as well as in increased productivity from use of more mass transit or telecommuting. This could be done by establishing goals and a vision for each of the key environmental sustainability measures with teams using a quality approach supported by a process owner for each. For shore water quality our goal is zero beach-block-days of closures due to pollution and our vision is a small child playing at the water's edge. We also liked New Jersey residents being able to see their toes in the water—an improvement they could see! This outcome, which provides an improvement that citizens can see, is what will help improve the overall perception of New Jersey as a good place to live.

The process owners for the shore water quality were the 90 shore municipalities. New York City could be considered as the 91st shore municipality that was also a process owner with "home rule" authority. New York City also needed special leadership, support and motivation that might be established by expanding our team to be a Quality New Jersey–Quality New York environment team, which would need the support of both the New Jersey and New York DEP agencies. These municipalities had home rule authority and responsibility for prevention of the major sources of shore pollution from sewage and storm-drain systems. While others contributed to the shore pollution problem, including farmers and boaters, their contribution was a distant second to that from the sewage and storm-drain system.

Our QNJ environment team supported the New Jersey DEP in leading, supporting and motivating those New Jersey shore municipality process owners. Help in the support effort was also received from the New York–New Jersey regional EPA, the four shore counties, the five shore watershed regions, and a number of shore environment commissions and environment associations.

Process owners need to be identified for each of the remaining sustainability indicators. For example, the process owners for energy and fuel production are utilities and oil companies. The process owners for energy and fuel consumption are New Jersey citizens. They are the ones that must be led, supported, and motivated to make continuous improvements.

Measures by themselves simply tell you where you are. In fact, representatives at the conference complained that, while a New

Jersey state plan for the environment had been developed in 1987, little progress had been made on most key indicators identified at that time, such as farmland protection. The primary cause of continued loss of farmland was the economic incentive for the farmland property owners to sell their property to developers for economic gain, and the lack of municipal commitment to retain a farmland zoning requirement on existing farmland.

In 1954, the words "Garden State" were approved for the license plates in New Jersey. Since that time, farmland acres have fallen by 50 percent, from 1,600,000 acres to 840,000 acres in 1997. In 1997, New Jersey has 9200 farms and the average Jersey farm is family-owned and about 100 acres. New Jersey Agriculture Secretary Arthur R. Brown, Jr. says that New Jersey has the highest per-acre agricultural land value and land taxes in the country. It would make sense that if we wanted to avoid a continued reduction in our New Jersey farmland, to protect the bit of "garden state" we have left, we should stop taxing it at the highest rate in the country and require that it remain zoned for country use. While such a tax and zoning requirement at the state level would infringe on municipal home rule authority, it would be in line with the benchmark approach taken in the Netherlands to protect their countryside.

Government changes such as these will be required in order to protect our environmental assets.

Similar to the shore water quality and farmland, each of the sustainability indicators could be improved by an analysis of causes, and by the development and deployment of appropriate countermeasures. The analysis of causes and the development of countermeasures is a relatively simple job that expert planners can do. The support for the deployment of those countermeasures by the appropriate process owners is much more difficult and requires leadership at both citizen and multiple levels of government. Leadership requires the following:

1. A vision
2. Clear goals, objectives and measures
3. Support for the analysis of root causes
4. Development and support for deployment of appropriate countermeasures

5. Incentives, both positive and negative, required to motivate deployment

6. Frequent clear communications.

In addition, the leaders must keep their boots on when the going gets tough and citizen leaders must maintain the continuous improvement effort and partnership with government over the long term even as government leaders come and go from both political parties.

The goal of achieving a sustainable economy, which neither New Jersey nor the rest of the United States currently has, is a desirable goal for the nation, for New Jersey, and for each municipality. This goal is beyond just balancing the budget. It is a goal for doing something useful with the budget. It is a goal that avoids depleting valuable environmental assets that are not accounted for in the government's financial reports.

It's the Economy *and* the Environment, Stupid!

When President Clinton was first elected in 1992, the campaign's reminder of what the public cared about was: "It's the economy, stupid!" In future campaigns it would be nice to see the reminder of what the public cared about to change to read: "It's the economy *and* the environment, stupid!" It's the income statement *and* the balance sheet! It's growing our GDP *and* protecting our valuable environmental resources!

The goal of achieving a sustainable economy protects today's environment and future generations from today's generation using up their environmental assets and leaving behind clean-up costs and pollution prevention costs. In some cases (for example, the greenhouse gases), we do not even know how to clean up the pollution of space once we pollute it. Therefore, we must place an even greater urgency on preventing continued pollution now.

A sustainable economy leaves the gardens green and the oceans blue for this and the next generation, while keeping business in the black. Accomplishing this may require cutting some government red tape along the way, by changes in our governance system.

LEADERS' VIEWS ON SUSTAINING CONTINUOUS IMPROVEMENT

I asked a few leaders whom I respect, and have had the pleasure to work with, for their views on what is essential to attaining a culture of continuous improvement. I also asked for their views on what is essential to sustaining it through leadership changes. Their opinions follow.

Bob Allen, former Chairman and CEO of AT&T

The major factor that impacted AT&T's cultural shift toward continuous improvement was the recognition that we had to commit ourselves to a quality process that would be sustained over time; one that we came to understand was really connected to business performance; and one that recognized that in a competitive world, "treading water" means falling behind.

While senior leadership is absolutely essential to improving process quality, customer satisfaction and improved business performance sustainability is only possible if the front line people truly understand and believe in the quality tools as a means to enhancing top line growth (or share) and bottom line profitability. Thus, moving from the theoretical level of quality to the practical is essential.

Further, a values statement that infers high standards of quality and continuous improvement will help sustain their practice. Employees must believe in their pursuit and constantly must endeavor to relate those values to their daily tasks.

With respect to maintaining continuous improvement, even with changes in leadership, one would hope that quality principles and practices are so ingrained in our behavior that a leadership change would not negatively impact our continuous improvement efforts. However, in the passing of the baton—or actually before passing—it may be important to have visible and audible actions and statements that reinforce their importance and demonstrate commitment to their continuation.

In our case at AT&T, I believe that the reaffirmation of our values—and the implied reaffirmation of our focus on continuous improvement in an even more competitive environment—has

Table 10-1.

Role of AT&T Chairman and President as Quality Leaders (1997)	
Chairman & CEO	**President & COO**
• Continue to champion quality and customer satisfaction • Charter new chairman's business system assessment process (CQA) • Continue as Malcolm Baldrige Award Foundation director • Personally recognize achievements	• Communicate explicit quality and customer satisfaction goals and objectives • Lead president's reviews of business system assessment (CQA) • Continue review of customer satisfaction at quarterly results meeting • Lead quality assessment of AT&T-level management system (1998)

already served to maintain a culture bias. In fact, the expressed intent to be even more customer oriented carries with it the need to continuously know and serve our customers better (Table 10-1).

Frank Ianna, AT&T Executive Vice President, Network and Computing Services and Chief Quality Officer

The absolute key to achieving continuous improvement through a quality approach is proving that it really works. The quality improvement effort needs to be clearly linked to results on an important business issue. When employees understand the connection between using a quality approach and obtaining improved business results it encourages use of a quality approach to obtain continuous improvement in that area, as well as in other areas of the business. Communication about the business success also needs to acknowledge that the quality approach was a major cause of the successful improvement effort. This usually requires the team, and leadership of the team, to openly acknowledge and recognize the use and importance of the quality approach to the success.

Replicating improvement efforts in other parts of the organization is required for continuous improvement. The key to replication of improvement efforts is to understand and encourage the use of the underlying quality tools and methods that enable replication by many of what a few have done.

In order to sustain continuous improvement when management changes, the organization should stay focused on what the customer wants instead of changing focus to what the new boss wants. The organization needs to continue improvement efforts on behalf of the customer. However, when significant market or business changes occur, the quality approach to supporting continuous improvement may need to be refreshed to address a new business situation or to support a new organization. The quality approach to support continuous improvement of an organization should change as appropriate to support changes in strategy that are driven by the marketplace.

Finally, the individuals in the company must know that the company really cares about continuous improvement and that the company supports a quality approach to achieving that continuous improvement, because it works (Table 10-2).

Table 10-2.

Role of Unit Heads as Quality Leaders
1. Ensure that unit's strategic and operational plans are aligned with market-facing units to support execution of end-to-end, bundled strategy.   2. Communicate explicit quality and customer satisfaction goals and objectives.   3. Support use of a quality approach throughout the unit, that is, PDCA: *PLAN*—Ensure goals are set using benchmarks and policy deployment   *DO*—Ensure key processes use best practices   *CHECK*—CVA/CQA   *ACT*—Use problem-solving to close performance gaps   4. Review key business process improvement plans and ensure resources are allocated appropriately.   5. Review of customer satisfaction and quality results at results meetings.   6. Use of recognition and rewards to motivate continuous improvement.

Table 10-3.

Role of AT&T's Quality Director

Assist unit in achieving a culture of continuous improvement by:

- Developing and implementing plans to attain CVA objectives
- Leading the unit's participation in CQA initiatives, with emphasis on improvement plans to drive results
- Developing and implementing plans to deploy AT&T's quality approach (PDCA) throughout the unit, including benchmarking, policy deployment, process management, problem solving, and CQA
- Developing and implementing plans to utilize best practices throughout the unit

Laurie Groves, QNJ Environment and AT&T
Quality Office Team Member

Continuous improvement requires several things (see Table 10-3):

- Partnerships
- A team
- A team leader to stay organized
- Team fun and team results to motivate continuation
- Rewards and recognition
- Awareness of the difference an individual can make
- Time commitment

Judy Soltis, QNJ Environment and Lucent Technologies
Quality Office Team Member

Continuous improvement requires a focus on the end customer at all times. The end customer sets your goals. When you think you have achieved excellence, just ask your customer to set your next goal. The end customer will have no problem in making it clear that continuous improvement is always required.

When a leadership change occurs just keep on going. Don't stop to wait for the new leader, who will need time to learn about the new organization. When he or she gets to you, be ready to sell the new leader on the continuous improvement effort you have been focused on for the customer, showing results obtained and

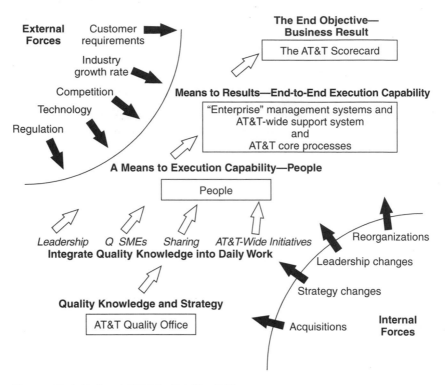

Figure 10-1. Role of AT&T's Quality Office

goals for the future. Most leaders are in favor of continuous improvement for the customer, so if that is what you have been doing, you should be supported to continue (see Figure 10-1).

John DiBiaggio, President of Tufts University. John DiBiaggio addresses the importance of education in establishing an early focus on the citizen's responsibility to contribute to continuous improvement and the need to learn how to contribute to continuous improvement. The following is from an article he wrote for the Boston Globe, *published on May 2, 1997, providing a view from local university presidents, in "The President's Corner" of the paper:*

The truth is that without passion for a cause, for an issue, there can be no lasting sense of responsibility. Neither can there be meaningful action without an ability to analyze and decipher the conflicting complexities of every issue a free society faces. And this is where our tremendous responsibility as educators comes into play. The most important skill we can teach is that of balance—between passion and pragmatism, between philosophy and political reality. The one hope for preserving a democracy is to reempower our students with a sense of civic duty, where knowledge is not intellectual elitism, but the ability to work for justice and prosperity. It lies with us to change America's current peril into possibility.

Wendy Rayner, New Jersey DEP Chief of Staff

One of the keys to attaining a culture of continuous improvement is to educate leadership about the principles and benefits possible from changing how the organization conducts its business in order to cope with new, challenging issues. The CEO and executive-level managers must understand that change requires creating new systems, and new systems, in turn, demand leadership. Training leadership about their new role in leading the change is the first step.

Change is impossible unless the head of the organization and its key managers actively support the continuous improvement process. That is their primary role. They must "walk the talk." Leaders need to consciously become living symbols of the new culture.

Communication is another key factor in the success of changing an organization's culture. You cannot overcommunicate the principles, concepts, and the successes of change throughout the process. Truly effective communications encompasses both words and deeds—with deeds being the most powerful.

Another key is to create a shared vision that change-management is important and urgently required in order to make necessary improvements and progress. Involve as many employees as possible in the process. Changing a corporate culture takes a long time and employees need "short-term wins" in order to support the new system. Short-term goals should be set that can be easily communicated and easily met in order to keep employees invested in the new process. This is picking the low-hanging fruit first.

Both short-term and long-term goals should be set, along with the appropriate performance measures, in the yearly planning process. These performance measures will not only provide direction, but also indicate how much progress is being made toward new goals. In order to anchor the changes that develop within the culture, employees should be recognized for how their new approaches, behaviors, and attitudes have helped improve performance. To sustain their efforts, all levels of the organization need to see continued progress toward the goal of the organization.

Finally, the leadership needs to be sure that the next generation of managers embodies the new approach. Promotions need to be tied to performance and support of the culture of continuous improvement in order to get the culture change to last.

Dave Rosenblatt, New Jersey DEP Coastal Monitoring Supervisor

A culture change with origins in the extensive pollution incidents of the mid-1980s has occurred for those of us in local, state, and federal agencies involved in coastal water-quality issues. A significant factor contributing to this change has been a goal of zero beach closings, set to a large degree by the expectations of the public. I have pursued this goal by directing the attention of the appropriate permitting and compliance agencies to the issues, and by gaining their support for attaining the goal. To maintain the culture of continuous improvement in the shore water improvement efforts during leadership transitions, only results will do.

It is too early to say that the entire New Jersey DEP has achieved a culture of continuous improvement. For the culture to change, clear meaningful goals must be set and accepted by those who would attain them. We have started but not yet achieved a culture of continuous improvement on a department-wide basis. To do so would also require team (municipal, county, state, and federal) efforts and successes in the areas of land, drinking water, air, streams, and river water quality.

A culture of continuous improvement will be maintained through leadership transitions when successes in process improvements are observable, and the staff recognize that they, themselves, are the reason for the success. When the staff is fully engaged in a quality system, new leadership will have no reason to interfere with it. Of course, without a fully engaged staff, the quality system has a tenuous existence.

Nick Finamore, AT&T Loaned Executive to the
New Jersey DEP (1996 & 1997)

There are a couple of key factors that have become evident as we attempt to change the culture at the New Jersey DEP to one of continuous improvement. We are starting to see examples of success and are beginning to understand what it takes to succeed in government, but we are very early in our journey to change the culture. First and foremost, it is my observation that we need the shared understanding and active involvement for the quality initiative from all of the senior leadership team, with visible and ongoing application to their own day-to-day management activities not unlike what we expect to see in a business environment. Generally in government, the top leadership team is appointed from legislative and/or legal positions, where they have worked primarily as individual contributors. For the most part, they have not had the benefit of leading large organizations and, thus, are placed into situations in which training in management and leadership skills is essential, with a focus on managing with continuous improvement in mind.

At the turnover of a new administration it is imperative that the new leadership team be educated together in the value of either continuing, if the program is already in place, or introducing these principles into their operations.

Many of the senior leadership team express concern that these principles will not work in a government arena because of the political influences that occur day-to-day. Top management support appears to be necessary, but not sufficient to carry the load. Career middle managers and staff also need to be convinced of what is in it for them.

We are introducing continuous improvement activities at a time when government is downsizing and, therefore, many view continuous improvement as a way to reduce the work force. Our biggest challenge has been to convince people that this process gives them more participation in decision making, and that they can make their job more enjoyable by ridding their operation of inefficiencies. One of the primary reasons most people have come to the DEP agency is to improve the environment. This is a key shared value for the department, and one that a majority can relate to.

Success appears to come when cross-functional process teams are equipped with the appropriate quality tools and techniques. They find improvement opportunities themselves in core operations and become convinced that the quality approach does not necessarily result in a reduction in work force, but can allow them greater control of their operation leading ultimately to a greater environmental benefit.

One of the greatest challenges is to link day-to-day activities with environmental goals of the agency. The key factors to success at the DEP appear to be convincing both top management and middle management and the staff closest to the action, of the value of using these continuous improvement principles to achieve improved environmental results.

Lester Jargowsky, Monmouth County Health Department Health Officer

To obtain a culture of continuous improvement we need a core capability or expertise in how to improve the environment to meet health requirements. The County Health Department provides that expertise for the area it serves.

We needed volunteers to apply that expertise in their communities. This includes the municipalities, the environmental commissions, the environmental associations, the marina trade association, and the Coast Guard.

Third, we need partnerships like the Monmouth–Ocean County Alliance which includes a 17-member watershed planning group to improve the Manasquan River water quality. This group includes the New Jersey DEP, counties, municipalities, and environment commissions and associations. As a result of four-and-a-half years of work, significant improvement in water quality has been made.

We have seen a significant improvement in the value of our environment by the public over the past twenty years, thanks to the work of our schools in the education of our children and the work of our environment associations in the education of the public. Twenty years ago, after a Fourth of July holiday, we would have beer cans everywhere. This year, I know after a holiday the area will be litter-free, because people care about their shore and now take care of it.

We need the sustained commitment and resource support from government. In the case of the county that means commitment from the freeholders. After the 1988 shore pollution crisis we got the commitment needed to improve our shore water quality and the freeholders have sustained that commitment, although each year the money gets tighter. To sustain that commitment we must continue to show the connection between good water quality and good business. The freeholders listen to the public; therefore, if you want support from the government, you must get support from the public. The public must be educated on the economic benefits of clean water quality, and see the money spent on it as a good investment that pays off for the community—and makes New Jersey a nicer place to live. The QNJ award also made a very positive impact on my freeholder.

To sustain the culture of continuous improvement when leadership changes, we should be very careful of who we select as our new leaders. In addition, we need to keep the public on the side of continuous improvement in the environment to be sure that the public will keep politicians who will keep the government workers who will support and partner with the volunteers.

The dolphins are very visible. We now have whales 17 miles offshore on a regular basis, and striped bass are running strong. The return of this type of marine life is what gains public support for continuing our water quality improvement efforts—and makes this job very worthwhile.

Martin Pagliughi, Mayor of Avalon, New Jersey

The wake-up call was the death of the dolphins in 1987 and 1988. We had waste washing up on our beaches. The state and counties closed beaches due to pollution, which caused a drop in our tourism business. The Jersey shore's economy is based on tourism.

We put together a storm-water management plan that included continuous improvements on a year-to-year basis. We planned to remove or divert one outflow pipe per year until all the pipes that were contributing to beach pollution had been remedied. We committed to step-by-step improvements that we could afford to do each year—and then we did it.

In 1994, we established a management contract with New Jersey American Water Company to continue our maintenance and improvement plans for our drinking water, storm water, and sewage water systems. In 1997, we improved the storm-water pipe system that feeds into the bay, to reduce potential pollution on the bay side of the island.

Providing good drinking water and proper disposal of storm and sewage water are major responsibilities of community governments. Because continuous improvement programs must be funded, we plan the annual funding of water-related improvements as a key part of our budget.

Avalon is an island only five miles long and a half-mile wide. However, our taxable property is valued at $1.6 billion, which is the source of our county property tax revenue. Our job is to ensure that the community's property values are maintained and improved, which requires taking good care of all of our water systems. When we successfully manage our on-shore water systems, we also take good care of the ocean, which in turn takes good care of our tourism business, which in turn takes good care of our property values, which in turn provides the basis for tax revenue, which in turn allows us to manage our water systems, thereby closing the loop.

Our annual property tax revenues of $14 million are less than 1 percent of the taxable value of the property. Half of our community taxes go to pay for shared services, such as schools and libraries, provided by the county. Of the remaining community budget, $4 million goes toward water system-related expenses. Of the $4 million, two-thirds goes to expenses in the sewage treatment plan, including necessary plant and process improvements. The remaining one-third covers our contract with New Jersey American Water Company for maintenance and improvement of our drinking water, and storm-drain and sewage systems.

An annual capital investment in our water systems is part of our continuous improvement focus, which we have accomplished while maintaining a stable level of debt at one-third the maximum allowable for our community.

Unfortunately, some communities have let their water systems degrade because annual maintenance and improvements are not visible to the taxpayers. After several years of neglect, the cost of repair and replacement becomes very expensive, and the community is faced with significant water problems. Continuous improvement is necessary to maintain our water quality and to sustain a reasonable affordability level on an annual basis.

It is the responsibility of a community's leadership to communicate to its citizens the need for annual improvements in the water system and the value of sustaining the continuous improvement effort. Maintaining funding support for continuous improvement requires constant education on the value of continuous improvement.

If we have a community that is educated on the value of continuous improvement, then even when leadership changes, the taxpayers should expect continuous improvement to be part of any new leader's plans.

1997 SHORE SEASON KICK-OFF

For the past seven years I had sent the shore mayors and county health officers a letter, at the opening and the closing of the summer season, reminding them of our shore water quality goal, the primary causes of pollution, and our past accomplishments. Included was recognition of those shore towns and counties that had won the QNJ Shore Quality Award in prior years, and encouragement for applications in the current year.

The 1997 letter was ready to be sent by New Jersey's commerce commissioner Gil Medina and me. We told the mayors that shore tourism revenue was up again in 1996, and we had achieved two consecutive years of great shore-water quality, 1995 and 1996. Both of these results were due to the continuous improvement efforts of the shore municipalities, supported by the shore counties and the state DEP. This year, we also pointed to the role our governor and DEP commissioner were playing in leading a coastal partnership, including environmental associations, in developing a plan for further shore improvements.

We asked the municipalities to expend continuous effort to prevent pollution and to quickly repair problems when pollution occurred. The letter also cited the names of the six winners of the

1996 QNJ Shore Quality Award. The letter asked each town to set a goal of zero beach-block-day closings in 1997 and to apply for the 1997 QNJ Shore Quality Award. The 1997 awards would be presented in November at a QNJ quality conference. (See Appendix B for the shore letter.)

It was just a few weeks before Memorial Day, the official beginning of the summer shore season. Our QNJ environment team met to continue our work on the shore water quality improvement efforts for the ninth year. We had invited the New York DEP to our May team meeting to discuss how New York might also begin using a quality approach and become part of our extended QNJ team, allowing us to address together common bay and ocean water-quality issues. In the short-term, we wanted to help ensure that New York developed a plan to deal with their pumping station repair in a way that would prevent dumping raw sewage into New York–New Jersey waterways.

In addition, a major topic for the May meeting was discussion of a plan to eliminate all the combined sewer overflows (CSOs) on the New Jersey side of the Hudson River to prevent overflow of raw sewage from New Jersey during heavy rainstorms. New Jersey needed to become a role model for New York by leading in cleaning up the CSO problem on our side of the river in order to prod New York into doing the same on their side. As long as New Jersey neglected to clean up our sources of raw sewage overflows, New York had an excuse to do the same on their side of the river, even though New York had a much larger volume. We knew prevention would require a large capital investment. New Jersey had already made a lot of the less costly improvements, the "low hanging fruit," in the sewage system. Of course, picking low hanging fruit in a sewer is not exactly fun or easy. Continuous improvement would now require addressing the outdated combined sewer overflow (CSO) problem. New Jersey accounted for about 22 percent of the raw sewage that overflowed through the CSOs into our New York–New Jersey waterways, with New York contributing the remainder.

With the help of AT&T's public relations department, I also prepared a news release in May for the 35 shore weekly newspapers that I hoped would make their issue for the opening of the summer season. At the opening of the summer season, people on the shore are always concerned about forecasts for the weather, water quality, and tourism business. My news release made no promises on the weather, but predicted good shore water quality again and increased tourism for the summer season of 1997.

I felt this prediction was safe to make based on the culture of continuous improvement and teamwork by the shore municipalities, shore counties, and the state DEP. Our QNJ environment focus group had helped to achieve and sustain this culture over the past nine years using analyses of data, quality tools, and methods to understand root causes and develop appropriate solutions. The team supported others in using quality principles for prevention and fast repair, setting goals and objectives, and measuring and reporting results. Also, we provided coaching and guidance, positive recognition and sharing of best practices, to encourage and speed replication of successes (Figure 10-2).

Our New Jersey shore partnership had developed the culture of continuous improvement required to sustain a process capability

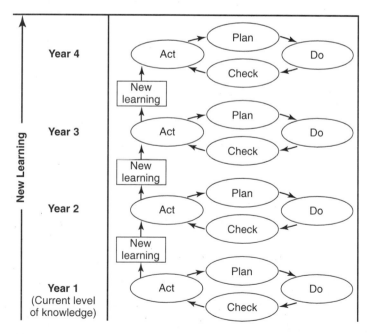

Figure 10-2. The Quality Spiral

that could prevent pollution and/or make fast repairs in the case of a pollution problem. As a result of the consecutive years of great shore water quality, tourists had renewed confidence in the Jersey shore. Summer rentals were sold out before the summer season even began, again ensuring a great tourism season.

Our QNJ environment team had to continue to sustain this culture of continuous improvement, to set new goals and objectives, to focus on new problems and to continue the recognition that motivates the continuous improvement required to prevent old problems from recurring. Our focus group was committed to creating and sustaining a culture of teamwork and continuous improvement. We had created it, but sustaining it required continued support of the paradigm of goals, support, and recognition. For example, development of a plan to replace the New Jersey CSOs and the expansion of our team, and our goals, to include New York City were important next steps to support the continuous shore improvement effort we had started.

On Thursday, June 19, 1997, only a few weeks into the summer season, the *New York Times* reported that a New York sewage spill in Westchester County closed all 26 of its beaches on Long Island Sound because of the lingering effects of a 100,000-gallon sewage spill early in the week, near City Island. The spill began shortly after 3 P.M. on Sunday when a pipe that carries raw sewage between the Bronx and City Island developed a crack beneath Eastchester Bay. Officials said beaches might not open for the next weekend.

One of the beach areas affected was in the city of Rye. Mayor Ted Dunn said, "The sewer systems all along the Sound are old and we've been talking for years about upgrading them, but all we've done is talk. We need to upgrade the pipes that run underground and upgrade the sewage plants themselves."

It appeared the winds and tide would carry this spill toward Connecticut. In addition, in Greenwich, Connecticut, three beaches were closed on June 18, 1997 because of an accidental discharge of sewage from a sewage treatment plant near Byram Shore Beach. Byram Shore Beach is several miles north of Rye, New York, with the Westchester Country Club between them.

Our QNJ environment team had not repaired any sewer plants, fixed any sewer pipes or picked up floatables in the New York–New

Jersey harbor. However, we had helped develop and support a culture for continuous improvement along the Jersey shore and had achieved and sustained the most improved ocean water quality in the nation. Developing and sustaining that culture for continuous improvement is a real job that has a real impact on results. Prevention requires a culture of continuous improvement by all the people in the process, or the pollution will return. Since it is the dolphins that we wanted to return each year, not the pollution, we must continue to sustain that culture of continuous improvement by involving everyone in the process of keeping our New Jersey shore water clean.

Going forward, we need not only to sustain the shore water continuous improvement effort on the Jersey shore, but also expand that culture of continuous improvement to the surrounding states and to the other key environment areas in New Jersey.

Lessons Learned (1987–1997)

The following summary is grouped into five areas, taken from earlier chronological chapters: Business and Government Leadership, Team Leadership, Problems and Causes, Continuous Improvement, and Government Issues.

1. **Business and Government Leadership**
   - The customer can provide a clear and compelling goal that the team cannot.

   - World competition requires world class standards of excellence.

   - The gap between our current performance and world class must be clearly communicated, and the expectation set that it will be closed by a set date.

   - Improved quality produces improved customer satisfaction, which results in increased revenues.

   - Clear responsibility and accountability for measurable results and the process that produces the results is required.

   - Ownership for the operations and support processes needs to be clearly defined.

- Support processes need to include both recognition and enforcement for the operational processes.
- Those controlling the resources must be informed and supportive of the improvement efforts.
- Recognition, combined with financial support, can accelerate replication of best practices and the rate of improvement.
- The Japanese have a higher standard for senior management's knowledge of their quality approach.
- A visible improvement is required to satisfy customers.
- Quality improvements should also produce cost reductions.
- Quality excellence produces business excellence for all stakeholders.
- Excellence must be led and supported to ensure the effort includes a critical mass.
- A role-model example alone is not sufficient—training and management support are required to replicate success.
- Withdrawal of financial support for best practices use results in less use of best practices.
- Reorganizations should be done to support execution capability of processes, people, and relationships. It is not strategy, then structure; it is strategy, processes for execution of strategy, then structure.

2. **Team Leadership**
    - It is tempting to blame someone you don't trust for being the cause of serious problems.
    - We can learn from the work of many people, from a variety of places.
    - Experiential training can provide a sound accelerated start for a problem-solving team.
    - Trust is required, and must be honored, when data are shared. Data must not be used to blame, but to find root causes.
    - People will try to beat the measures until you begin helping them beat the root causes of the problems.
    - Teamwork requires measurable team goals.
    - Excellence is a stronger motivational goal than setting minimal standards.
    - Positive recognition can motivate people to excellence.

- Giving recognition can be more fun than giving fines, and more helpful.
- Punishment does not ensure minimal standards will be met.
- Setbacks will happen. Keep your boots on when they do.
- More people working on a solution means more solutions.
- People must know how much you care, before they care how much you know.

3. **Problems and Causes**
   - It is tempting to blame what is visible for all the problems.
   - The causes of a serious problem are not always easily visible.
   - The actual causes of a serious problem are usually multiple in number, and difficult to uncover.
   - Twenty percent of the root causes are responsible for 80 percent of the defects. It is critical to know which 20 percent, using data not emotion, and focus on it.
   - Once you find all the root causes, the solutions to the root causes are usually obvious.
   - After solving a large problem, you're likely to discover smaller problems.
   - Root causes of problems and solutions are required to prevent or reduce problems. Cleanup alone isn't sufficient to prevent reoccurrence.
   - Controlling problems at the source is the most efficient and effective means to reduce cost and improve quality.

4. **Continuous Improvement**
   - Some improvement will result from reacting to disasters, but reaction is not sufficient to prevent disasters.
   - We need more than one role model.
   - Sharing local solutions is a useful role for support staffs.
   - The quality journey is not without setbacks.
   - You need to stay focused on your goal and eliminate the sources of the problems that affect reaching your goal.
   - A continuation of effort is required to sustain the gain.
   - A fast repair strategy is required to achieve minimum performance standards, and a root cause prevention strategy is required to achieve excellence.

- Continuous improvement requires continuous discovery, continuous development, and continuous maintenance.
- Measures of results (samples) are required to provide data for control and improvement.
- Without measures you can't develop appropriate countermeasures.
- Proper analysis of data can provide key information to guide the direction of improvements.
- Data is required to prioritize improvements needed, given limited time and money.
- Instead of a choice between either/or, more often than not, both are required.
- The Malcolm Baldrige National Quality Award criteria can be used to check on progress, provide feedback for improvement, and recognize excellence.

5. **Government Issues**
   - Our governments rely too heavily on laws, regulations, and punishment.
   - Punishment does not accelerate improvement.
   - Adjusting to lower standards does not motivate achievement of excellence.
   - Citizens must be involved to help set goals for excellence for our society. They are the customers for government services.
   - The governance process, as well as key operational processes, in business or government, has to be continuously improved to meet changing marketplace situations and new stakeholder requirements.
   - National goals and local solutions work better than no goals and national solutions.
   - Similar problems exist throughout the nation. We need to learn from successful solutions and support their replication.
   - Partners require a common goal and a common approach.
   - Partners don't sue each other.
   - Government doesn't regulate itself very well.
   - A cost/benefit analysis should be used to prioritize major improvements.
   - To maintain our current standard of living and our future standard of living, we need to improve at a faster rate.

- We need a quality education, and a quality approach in our education.
- Taking care of our infrastructure is taking care of our country.
- Information technology used in business can be applied to help solve problems in government.
- A sustainable growth plan requires goals, rules, and resources for improvement.
- Our government should not require subsidies for things citizens do not want and should fund things that support our objectives.
- The municipalities, with home rule, may be the process owners for many key areas of government.
- The federal, state, and county levels of government need to lead (see Business and Government Leadership above) and support the municipalities in their continuous improvement efforts.

CHAPTER ELEVEN

Continuous Improvement of the Governance Process

(The Future)

JULY FOURTH, 1997

July Fourth is my daughter Michelle's birthday, and I always have a barbecue for her on her birthday. Michelle turned 23 on July 4, 1997, and she came over for the barbecue with her five-year-old son, Trevor. Since it was July Fourth we read the Declaration of Independence, and it gave me a few ideas for a Declaration of Continuous Improvement.

Declaration of Continuous Improvement—of our Governance Process

Our government was established by the people and secures its powers from the people.

Our government was established to secure certain rights, such as life, liberty, the pursuit of happiness, and promotion of the general welfare for today's citizens, and for our next generation.

When the governance process is unable to secure these rights, then it is the duty of the people to alter the governance process and establish an improved governance process that seems most likely to affect their general welfare, and that of the next generation.

It is self-evident that today's current governance process is increasing the liabilities that will be left to the next generation, thereby reducing their general welfare, as a result of the following:

1. The continued increase in the debt of the federal government, which is now over $5 trillion.
2. The continued national trade negative balance of payments.
3. The increasing federal entitlement program costs, which are projected to significantly increase in the next decade.
4. The lack of a sustainable development plan to protect our country's environmental assets, such as our national ocean water quality.

BARRIERS TO GOVERNMENT IMPROVEMENT

On April 28, 1997, officials from Vice President Gore's National Performance Review Federal Benchmarking Consortium Study on Performance Measures visited AT&T to benchmark our approach to measurement of performance.[1] This visit was part of a government study of several leading corporations and leading government agencies. The night before the meeting, I answered their site visit questionnaire provided to me on a floppy disk, allowing me to type answers on my computer. It took me three hours to answer the 22-page questionnaire. We spent another four hours the next day discussing my answers and how some business approaches might be applicable to government situations.

One of the documents they provided at the meeting was a list of 10 unique barriers that government faces in attempting to implement a performance measurement system. Since a big part of my role on my QNJ team had been to encourage government not only to measure performance but also to improve it, I asked if I could share their top 10 barriers in a book I was writing that also addressed how to implement quality in government. I was told that all government information could freely be used by anyone and they encouraged me to address the barriers they had identified.

The top 10 government barriers in measuring performance were identified as follows:

1. The federal budget process
2. The Paperwork Burden Reduction Act
3. Lack of accountability for results
4. Long lag-time between activities and outcomes
5. Limited knowledge of costs to achieve results

6. Focus on equity over efficiency
7. Constraints in hiring and training the talent required
8. Contracts awarded to lowest-cost vendors
9. Partnership consensus on measures required
10. Legislative direction, priorities, and funding control

These barriers to implementing quality improvement in government looked like the list I had been facing for the past several years with my QNJ environment team. This list is also similar to one that business people might come up with if asked what issues they faced in developing a performance measurement system.

Based on my QNJ environment team experience and my AT&T experience I decided to see if I could understand the causes and develop possible solutions to the barriers that are faced on a broader scale by federal government agencies trying to implement their quality improvement approach. I might also be able to apply the solutions to help improve my own efforts at QNJ and/or AT&T.

1. The Federal Budget Process

The perceived problem is the uncertainty of continuity in funding of programs and the average lag time from request to funding of almost three years. This is similar to the problem my QNJ environment team faced at the state level when the State Assembly took funding away from the ocean water monitoring program in 1992, without understanding that this program was required to support improvement of the entire process for improvement of the New Jersey shore water quality.

The cause of this perceived problem may be that the budget funds *programs* instead of processes or organizations that would be accountable for results.

The executive branch should have the flexibility to move money around to fund the programs required to achieve the objectives and goals set by Congress. Congress needs to establish performance goals and annual objectives and fund the agencies accountable for achieving those goals and objectives. Then Congress could empower the executive branch to make decisions on what programs to fund that are most likely to achieve the goals that Congress has set.

2. The Paperwork Burden Reduction Act

The perceived problem is that the Paperwork Burden Reduction Act places a paperwork and time-delay burden on getting the approval needed to collect the data required to measure performance.

The Paperwork Burden Reduction Act was probably well intended and aimed at making it difficult for the federal government to obtain individual personal information, as well as to reduce the burden of government-required data collection. Most people would agree with the intent of this act. However, improving performance requires having available measurement trends and data that can be acted on to improve results. This act could be amended to allow for the collection of performance data that was specifically related to the key congressional goals and objectives.

Until Congress modifies this act, I suggested use of electronic mail for their data collection requests, which could reduce paperwork to zero. When I responded to Vice President Gore's study, I answered the questions directly onto their floppy disk. Computer communications solutions to the rescue!

3. Lack of Accountability for Results

The perceived problem is that public agencies are being called upon to achieve outcome results that are beyond the scope of the programs they manage. Therefore, the agency can't actually be held accountable for the results of the entire process.

This was, and still is, the major problem faced by the QNJ environment team. We try to facilitate the appropriate resource allocation decisions in the existing decision-making process, involving four levels of government and three branches of government, using our analysis of the problem and the cost benefit of addressing them. We try to communicate our analysis and motivate continuous improvement by all involved, while managing the appropriate allocation of limited resources. We even try to share our analysis of the causes of problems, and the cost benefits of addressing them, with our upstream neighbors.

The cause of this perceived problem is the focus of the executive branch on managing programs rather than managing a process. Any process such as keeping the ocean clean uses a number of programs for improvement. The overall process needs to have an owner. Someone that assigns the available resources to accomplish the objectives. Someone that accepts overall accountability for the use of resources and responsibility for the achievement of results.

For example, we need Congress and the state assemblies to set goals for improved ocean water quality, and other key environmental areas. Executive branch agencies need to submit budgets to achieve continuous improvement toward those goals while developing means to reduce, or at least hold flat, the overall cost of their agencies.

The cause of the lack of accountability starts in Congress. Congress has maintained the responsibility for selecting and funding programs instead of setting performance goals and objectives. Congress is overmanaging and micromanaging the executive branch by managing programs instead of managing processes and outcomes. A reason for this is partly lack of knowledge on how best to manage for performance and partly Congress' desire to control the programs as a means to support special interests in members' own states. Whatever the reason, the approach to funding and managing performance at the federal level needs to be improved in order to get improved results.

Someone once said that insanity is believing that you will get different results by doing things the same way you always have. The funding process has to change to achieve improved results from government.

4. Long Lag-Time Between Activities and Outcomes

The perceived problem is that there is a long time-delay between program activities, such as a single education or health care program, and actual improvement in education or health outcomes.

Our QNJ environment team found that when we used the data from the process we were trying to improve and focused on the largest causes of the problems (a Pareto analysis), at any given time 20 percent of the causes accounted for 80 percent of the problems. When the executive branch has the authority to focus on the largest causes of the problem, progress can be made continuously as we did with the New Jersey shore water quality.

The cause of a very long delay between a single program activity and an outcome is that a single program by itself does not produce an improved process output. The process capability must be improved and controlled by managing a set of programs or activities that address the major causes of the problem.

To address the cause of this problem, the management approach of government should move from managing programs to managing processes. Processes may include sets of programs. The process manager should have accountability for the results and the responsibility to use a budgeted amount of money on the programs to provide the maximum improvement in the shortest interval. This approach will reduce the time needed to achieve improved results and avoid the concern about losing program funding. Program funding will be the responsibility of the process manager in the executive branch, while the responsibility for overall process funding and setting of goals should be with the process owners in Congress.

The roles, responsibilities, and accountability of both our executive branch and our legislative branch need to change to establish goals for continuous improvement and to enable management of the process to achieve those goals at a minimum cost.

5. Limited Knowledge of Costs to Achieve Results

The perceived problem is that many programs are implemented through state and local grants and therefore the program manager has limited information on the total costs to achieve the outcome results.

In the case of the New Jersey shore we had 90 shore municipalities and four counties spending money to achieve the continuous improvement results in shore water quality, along with some grants from the state, and state funding. The total cost and cost/benefit at the state level was understood by our team. The problem we had was the lack of government use of our cost/benefit analysis.

Information of the total costs, as well as process and outcome measures, should be a normal part of the expected information shared as a condition of receiving a grant. The cause of not having cost information is that it wasn't made a condition of the grant. Sources and uses of funds should be quantified and managed for each key process. No matter which level of government we look at, the source of money is typically the citizen who pays the taxes in one form or another.

6. Focus on Equity Over Efficiency

The perceived problem is that public programs must be more equity oriented than efficiency oriented. For example, efforts to increase efficiency cannot be done by sacrificing the equitable delivery of services.

In many cases, the term *equity* has been distorted to mean certain people get a piece of the pie. For example, in my work on the shore, the state grants to shore municipalities for mapping their sewer system the first year were given out based on miles of shoreline rather than focusing our limited resources on addressing key problem areas to obtain the best results in the shortest time frame. The second year, the program was cut and the most significant problem areas did not get help.

Equitable delivery of services means delivery in a just and fair, or impartial basis. Efficient delivery of services means acting or producing effectively with a minimum of waste or unnecessary effort. It would seem that both equitable and efficient services should be delivered. It is fair, just, and impartial to use limited resources to address the cause of problems, or the needs of the process for improvement, in order to produce the results that are best for the general welfare at the least cost. Simply giving those first in line an equal piece of the pie, and not eliminating the cause of the problem or making an improvement is neither equitable nor efficient.

7. Constraints in Hiring and Training the Talent Required

The perceived problem is the limitations imposed by the personnel process and the budget process to hire or train the people required.

Business has also faced this problem, and faced up to it. If the personnel process is limiting performance, it should be changed. The primary purpose of support processes, like personnel and finance, is to support the operational processes in achieving the desired outcomes at the least cost. If personnel processes do not support changing the people, through hiring or training, to deal with changing situations, then the people will fail. After they fail, the function may be outsourced to business, and the people will be changed without a chance to change themselves. The budget should allow the flexibility for managers to provide the training required for their people. People without skills doing a job is not a cost-effective solution, and not fair to the people in the jobs.

One cause of this perceived problem is the limited budget to train or hire the skilled people needed. The process owner should have the authority to use budget resources to produce the optimal results at the lowest cost, including the appropriate hiring and training of human resources in the process.

Another cause of this perceived problem is limitations on changing people in the process due to government personnel policies. Given the absolute need to improve government results and reduce government costs, human resource polices should be changed to support those objectives, not be a barrier to their achievement.

8. Contracts Awarded to Lowest-Cost Vendors

The perceived problem is that contracting the lowest-cost vendors does not help produce the highest quality performance measures.

Businesses have found that the highest quality supplier is usually the lowest overall cost supplier because quality suppliers reduce expensive rework costs by the purchaser. In other words, all the costs for all the steps in a process must be considered when selecting a supplier. The highest quality supplier usually does not have a significantly higher price and their higher quality usually saves significant costs downstream for the purchaser. A rule of thumb is that fixing a problem downstream costs 10 times as much as fixing it upstream. Hence, it usually makes sense to pay a penny more for quality upstream to avoid 10 cents to fix problems downstream.

The government needs to consider full-stream costs when making purchase decisions, as is now done in most businesses. This change in purchasing behavior to consider full-stream costs is one of the major reasons for the increase in importance of quality in business. Quality suppliers are now understood to be the low-cost full-stream suppliers. It is a mistake to only consider the first cost in the contracting decision.

9. Partnership Consensus on Measures Required

The perceived problem is that when programs involve grants, the measurements to be used must be arrived at by consensus of all the parties involved, and everyone involved wants a different measure.

The establishment of the overall process and outcome measure for New Jersey shore water quality of beach-block-days was a driver of other subprocess measures. For example, we created a measure of the completion of mapping of the sewer and storm drain system (a program) for each town and drove that by a measure of mapping completion dates. Mapping a sewer system does not improve water quality by itself, but it is a subprocess that enables fast repair cycle time, assuming the repair process (another subprocess) also has its subprocess measure, which was time to repair. Another subprocess is the repair and inspection of sewage treatment plants, with its own subprocess measures. Other subprocess measures include the cleaning of beaches daily by the 90 municipalities. A subprocess measure that our QNJ team still needs to establish is the replacement rate for combined sewer overflows (CSOs), which are a source of pollution each time we have a heavy rain.

Developing effective measures of performance is very difficult for a lot of reasons. The primary reason is usually lack of agreement on a common goal. As mentioned earlier, I believe that the government should establish a goal and appropriate measures prior to the awarding of a grant. A related problem is the confusion between process measures and outcome measures. The entire team needs to agree on the outcome, or results, measure, but several process measures may also be required to manage the subprocesses involved in the overall process. The subprocess manager may be able to decide on the appropriate measures without team consensus and simply obtain agreement from the overall process manager on the subprocess measure required to produce the best process output results.

Of course, if a single program is being managed by the grantee without outcome or process measures, it is not surprising that managers cannot come to agreement on a measure. A single program cannot have much impact and therefore the people involved will normally resist measurement of programs that they know can't have much impact by themselves.

Once we all know and understand the overall process outcome goal and measure, the subprocess managers can develop and manage their piece of the process and develop aligned subprocess measures that contribute to achievement of the overall goal. Before establishing a clear overall process goal for New Jersey shore water quality, we had subprocess or program managers blaming each other for the overall poor performance of the total process. One of our problems in enlisting help from New York is that they have not bought into our goal of improving New Jersey shore water quality. We probably need to expand our goal to include improving the New York–New Jersey ocean water quality to enlist New York as a full partner on our team—a next step for my QNJ-QNY environment team.

10. Legislative Direction, Priorities, and Funding Control

The perceived problem is that Congress provides specific program direction, priorities, and funding. The specificity of legislation on a program-by-program basis reduces the ability of the executive branch to make decisions that will influence positive outcomes for the overall process.

Congress should simply provide performance goals, objectives and process funding that encompass all the programs in a process. The role of Congress should change to that of overseeing outcomes and the improvement and control of the process required to achieve that outcome, instead of micro-managing programs that have no outcome measures. Instead of having program managers in Congress, we should have outcome and process owners.

QUALITY PARTNERSHIPS FOR EXCELLENCE

Minimum Standards

Government objectives are usually limited to ensuring that society meets minimum acceptable levels of performance. Unfortunately, our government is not a model for achieving excellence, even if those words may have appeared in a political speech or two. The focus of our government on minimal performance standards instead of standards for excellence is based on our fear of government control of the individual. To avoid being controlled by government, we decided to restrict government to a role of preventing society from falling below some defined unacceptable level of performance. This was a means to limit government involvement in the affairs of our country by those that prized individual freedom.

In fact, we also minimized the number of areas where government is allowed to define even a minimal performance standard.

Attempts to talk about addressing new areas that require setting of minimal standards are fought to avoid government expansion. Attempts to raise the minimal standards are also fought on the same grounds. We in the United States are generally concerned that any expansion of government will result in a loss of freedom for the individual, which leaves us with some tough questions. How can we achieve excellence in our society and not expand the role of government, or its control of our lives? How can we achieve excellence in our society and not give up our individual freedom?

The answer may be citizen and business involvement in our government instead of government involvement in our lives. Citizen and business involvement must not be for special interest reasons, but simply to work together as a team to solve community problems. The answer to achieving excellence, while not giving up our individual freedoms, could be citizen–business–government teams using a quality approach that is fact-based.

Given the role of preventing unacceptable performance, in the minimal number of required areas, the usual government approach to community problems relies upon a management approach that is driven by laws and regulations.

First, laws and regulations are created by a few very smart people. Second, monitoring of conformance to these rules is assigned to many who police and inspect in the community. Finally, we punish those we catch not following the rules. This legal approach to management is intended to achieve minimum performance levels by the majority, using negative incentives. Of course this approach has a very high cost in inspectors, police, lawyers, courts, and prisons. This approach also accepts the fact that a significant minority will choose not to meet the minimal standards, and instead choose to risk being caught and punished, which reduces the quality of this approach to below the minimal standards set, and increases the cost. Therefore, this is a high cost and low quality approach to governing.

The intent of this legal management approach is to motivate the majority of society to perform at some defined level of minimal performance. We can all get a D– on our performance, but if you get an F, and we catch you, there may be a punishment. The punishment is not certain if we catch you, because we also have to convict you in a court that requires evidence with proof beyond a reasonable doubt. There may be punishment for an F, but none for a D–. There is also no reward for getting an A or a B. Hence, a D– is the performance level that is motivated by this legal approach to management, and unfortunately many have accepted the F level as their life, with the risk of being caught and punished.

U.S. citizens are great at holding endless arguments about the criteria for an F or a D–. Why can't we set the standard a bit lower, or higher? Can you prove the world will fall apart, or that it won't, if we lower a standard? People seem able to accurately state the costs they would save, but not the additional cost to society if a standard is lowered. In addition, government is discouraged from providing minimal standards for new areas that need protection. For example, we still have no national standard, measurement, or goal for U.S. ocean water quality.

Quality Teams Strive for an "A"

In addressing the New Jersey shore quality problem our joint citizen–business–government team first asked the beachgoers what level of performance they wanted for their shore. How many days of beach closings did the beachgoers want? It turned out they wanted none. The beachgoers wanted an A level of performance on our valued Jersey shore. Therefore, our team set that as our goal, a goal that our government was unwilling, by itself, to establish.

During the summer of 1988, New Jersey had the worst shore pollution in the nation with 855 beach-block-days of closures due to pollution. We were at the F level of performance. However, neither the nation nor the state had even set a standard for what an F level of performance was. During the summer of 1997, New Jersey had 42 beach-block-days of closures due to pollution, a reduction of 95 percent.

With respect to our nation's shore water quality, the federal government (EPA) as of June 1997 had not even defined the minimum acceptable level for beach-block-days of closure due to pollution, although they have the power to do so. Neither the nation nor the states have set a goal for improving our national shore water quality. We didn't even have a goal defined for D– performance at the national or state level for our ocean water quality. However, the people in New Jersey knew we had an F in 1988, and it was unacceptable. People did not want a D– or to argue about what was an F and what was an acceptable, minimum performance D–. They wanted an A for their valued shore water quality. They wanted zero beach-block-day closings due to pollution.

Although the government does have a variety of regulations concerning industrial waste, the problem on the Jersey shore was mainly from sewage and storm-drain waste. Government was responsible for the sewage and storm-drain system, which was not under the same regulations as industry, and was not under control. Government does not do a very good job of regulating itself, which may be another good reason for citizen–business–government quality teams.

In fact, the chief of the EPA, Carol Browner, did a great job pushing through a higher standard for our national air quality, which was endorsed by President Clinton at the end of June 1997, to protect the public health. The air quality standards primarily affect industry, such as the power companies and the automobile companies. The national water quality standards we need would primarily affect the municipal governments that have not taken adequate care of their sewage and storm-drain infrastructure. We are doing better on our air quality standards than on our water quality standards. I believe this may be because government regulates business better than it regulates itself.

Improving Our Management Systems

To stay the best in a changing world, business and government need to continually adjust to the changes in the world and improve their management system's capability to address today's problems. A government system that was invented over 200 years ago was great for that time and for many years subsequent, but we should be working on continuous improvement of our government's management system approach, as required, to deal with today's problems. I think our forefathers would have expected that much from us.

Excellent quality calls for a different paradigm than the one government uses to meet minimal quality standards. To achieve excellence we need to make use of a lot of smart people who continuously learn from each other the best practices for achieving excellent results. To achieve minimum standards the focus has been more on conformance of many people to the best ideas of a few people. People can, and will, achieve excellent results when goals for excellence are set, new discoveries and knowledge transfer are supported, people are encouraged to add their own ideas on how to achieve excellence, and recognition for improvement and achievement is provided.

In business, and in government, we need both of the following:

1. Minimal standards below which negative consequences will occur.
2. Goals for excellence which, when achieved, will be recognized in a positive manner.

We also need partnerships across various levels of government to focus on the processes that need to be improved. An example of the partnership needed was announced on August 28, 1997 by the governors of New Jersey and New York. They joined with the federal EPA to announce a plan to restore and protect the New York–New Jersey harbor, the rivers and streams that empty into it, and the ocean waters stretching from Montauk Point, New York, to Cape May, New Jersey. This plan will include more than $6 billion to address problems caused by the combined sewer overflows in both states.[2]

SUMMARY

I had the pleasure of a great week at the Jersey shore in 1997—the best summer season in the past decade. My wife Jane, my son John, my daughter Michelle, and my grandson Trevor all had the opportunity to enjoy perfect ocean water quality.

The QNJ environment team used a quality approach to support the New Jersey DEP in making a major improvement in a process that is important to the citizens of New Jersey, the process to prevent pollution of our shore water. This process cut across four layers of government and required citizen involvement. The quality approach provided a nonadversarial approach for citizens to work within the system to make the changes necessary to drive significant improvement of the system. This approach can be replicated in other states and for other processes. In fact, other QNJ focus groups are working with New Jersey's government to address other key processes, such as education.

The QNJ environment team has agreed to expand its effort to address all of New Jersey's water quality issues by expanding our New Jersey Shore Quality Award and our quality approach to encompass all the New Jersey watershed associations. We have also invited the New York DEP to join our team to expand our quality approach and our goals to improvement of the joint New York–New Jersey rivers and harbors and the New York shore water quality.

In addition to this book, to aid others in the application of a total quality approach, a business case based on the preliminary manuscript has been developed for use in universities. This business case was developed by an AT&T–Rutgers graduate fellow, Stephen Pick, and the case can be obtained from Rutgers University. Additional information on the business case can be obtained at the Rutgers quality Web site, which also was developed by an AT&T–Rutgers graduate fellow. (The Web site address is: http://www.scils.rutgers.edu.qci/qci.html.)

The quality approach is an important tool that enables people to analyze problems from a fact-based viewpoint and encourage innovations and team-based solutions. The deployment of this approach also requires a culture of continuous improvement and continuous support of all the people making the required improvements.

The AT&T Quality Library

The AT&T Quality Library documents lessons learned in applying a total quality approach to improve business performance.

Leadership and Planning:

AT&T Total Quality Approach (500-452)
Batting 1000: Using Baldrige Feedback for Improvement (500-451)
Hitting the Target: AT&T PDCA Cycle (500-777)
Leading the Quality Initiative (500-441)
Policy Deployment: Setting the Direction for Change (500-453)
Quality Happens Through People (500-482)
Shared Values: Our Common Bond (500-479)

Customer Focus:

Achieving Customer Satisfaction (500-443)
Great Performances! The Best in Customer Satisfaction & Service (500-450)
Shared Expectations (500-794)

Supplier Partnership:

Supplier Quality Management: Foundations (500-496)
Supplier Quality Management: Selection & Qualification (500-497)
Supplier Quality Management: Monitoring & Improvement (500-498)

Process Management and Improvement:

Analyzing Business Process Data: The Looking Glass (500-445)
AT&T Problem-Solving Method (500-765)
Benchmarking Handbook (500-454)
Process Quality Management & Improvement Guidelines (500-049)
Reengineering Handbook (500-449)
Using ISO 9000 to Improve Business Processes (500-760)

To order AT&T Quality Library books, call **1-800-432-6600** or **1-317-322-6416**.

Award Letter to Shore Municipalities and Counties

**To: New Jersey Coastal Municipality Mayors
and County Health Officials**

New Jersey had the worst ocean water quality in the nation in 1988, based on a Natural Resources Defense Council (NRDC) study. Thanks to your efforts, New Jersey now has the most improved shore water quality in the nation. Shore tourism revenue has increased by almost $1 billion per year over the past few years as a result of our shore water quality improvement, creating the fastest growing sector of the New Jersey economy. In addition, the marine life along the shore are thriving, and the dolphins are back in full force.

While we have made a 90 percent reduction in our shore water pollution problems, we have not yet achieved our goal of zero beach-block-days of closures due to pollution. We need your continued prevention and improvement efforts to reach our goal and to avoid losing the gain we have made.

To recognize and share the best practices being implemented along our shore to prevent pollution, we will again invite your municipality, or county, to apply for this year's QNJ Shore Quality Award. This award has been established by Quality New Jersey's environment focus group, in partnership with the state of New Jersey, to recognize and share improvements in the quality of our shore water and beaches.

Each municipality and county that completes an award application will receive feedback on "strengths" and "areas for improvement" in their current efforts. In addition, we will share the best practices of applicants with all our shore municipalities and counties to help us move toward our goal of zero beach-block-days of closures due to pollution. A simple application form is attached.

Thank you for your work to date.

Gualberto Medina
*Commissioner of Commerce &
Economic Development*

Phillip Scanlan
*Chair, QNJ Environment Focus Group &
AT&T Quality Vice President*

1996 QNJ Shore Quality Award Application

I. **Describe improvement made in the following sub-processes of the storm water–sewer system needed to sustain clean water: (40%)**

   A. Storm water containment system
   B. Storm water disposal system
   C. Sewer infrastructure
   D. Sewage treatment plants
   E. "Upstream" (i.e., pipes), including those supported by the Sewage Infrastructure Improvement Act
   F. Replacement plan for old sewage and storm drain pipes, with plan for resources to fund the replacement required

II. **Describe improvements made in the following processes needed to sustain clean beaches: (25%)**

   A. Containment of trash
   B. Volunteer trash removal programs
   C. Street cleaning
   D. Upkeep of run-off basins
   E. Information management systems

III. **Provide results over a multiple-year period for the following key areas: (35%)**

   A. Water quality data: Beach-block-days of closures due to pollution
   B. Beach quality data

(The current QNJ Shore Quality Award application form can be obtained from the QNJ Web site by simply searching for the QNJ site.)

Shore Best Practices

1. Storm Drain and Sewage Infrastructure Improvements

Avalon

- Upgraded storm-water system in environmentally sensitive areas.
- Aerial photographs to assess flood mapping zone.
- Major projects for improvement of storm-water system.
- Inspection of all 600 manholes.
- Exploration of privatization of the water management systems.
- Mapping of the infrastructure onto a computer system to facilitate fast repair.

Asbury Park

- Installation of a new sewage treatment plant.
- Use of lakes for retention basins.
- Use of mesh frames on storm drains to prevent floatables from being discharged into the ocean.
- Inspection of commercial grease traps.

Brigantine

- Use of a vacuum truck to clean the inlets and pipes.
- Major investments in replacing sewer pipes.
- Replacement of deteriorated storm drain systems.

Manasquan

- Reduction in infiltration and exfiltration in the sewage system.
- Monitoring storm water before mixing with receiving waters.
- Daily mechanical sweeping of streets.
- Use of Clean Vessel Act grant money to make pump-out station improvements at marinas.

- Use of a sewer master plan to identify and improve problem areas.
- Provision of funds to upgrade the regional sewage treatment plant.

Margate

- Preventive maintenance of storm-water sewer system.
- Use of video camera in pipes to check for leaks.
- Installation of tide-check valves on storm-drain outfall pipes.
- Expansion of the inlet debris trapping system.

Stone Harbor

- Use of no chemicals in weed and insect control.
- Preventive maintenance of both stormwater and sewage systems.
- Use of grant funds and municipality funding to ensure proper maintenance of the storm drain and sewage infrastructure.

Stafford Township

- Removal of storm-drain pipes from the bay and terminating them in a natural aquifer that filters the storm water before it enters the bay.

Pine Beach

- Development of a five-year plan with four towns and use of DOT grant funds.

Monmouth County

- Repair of sewer lines feeding three major rivers.
- A proactive upstream estuary identification system.
- Closed an illegal dump in a marsh.
- Extensive use of an information management system and Web page.
- Partnering with Ocean County and municipalities to facilitate improvements.

2. Beach Improvements

Avalon

- Establishment of a bird sanctuary in Annacost Park.
- Commitment to beach renourishment.

Brigantine

- A trash containment program.
- Community volunteer programs.

Barnegat

- Aggressive approach to litter, trash patrol, and well-timed trash collection.

Stone Harbor

- Daily raking of beaches.
- Extensive volunteer and environmental programs
- Dune projects
- Bird sanctuary

Pine Beach

- Dog scoopers given to licensees.
- Boy Scout environmental projects and litter patrol.
- Sufficient placement of new trash receptacles.

1991–1997
Recipients of the New Jersey
QNJ Shore Quality Award

1997   Avalon
Stone Harbor
Avon-by-the-Sea
Northfield

1996   Avalon
Stone Harbor
Margate
Monmouth County Health Department
Pine Beach
North Wildwood

1995   Avalon
Asbury Park
Brigantine
Manasquan

1994   Avalon
Cape May County

1993   Avalon
Margate

1992   Stafford Township
Stone Harbor

1991   Avalon
Monmouth County Health Department

QNJ Environment Team Quality Cup Application
(1993)

1. Participating Team: Quality New Jersey (QNJ)
Environment Focus Group

2. Sponsoring Group: Quality New Jersey

3. Category: Nonprofit group 501C

4. Description of service:

Services: Assistance to the New Jersey State Department of Environmental Protection (DEP) in quality leadership, education, consulting and recognition on how to improve the New Jersey ocean water quality using a Total Quality Approach.

Clients: New Jersey State Department of Environmental Protection (DEP).

5. Brief description:

The team provided assistance to the New Jersey state DEP in applying Total Quality Management (TQM) to improve the New Jersey ocean-water quality from 1989 to 1993. In 1988, New Jersey had 855 beach-block-days of closures due to pollution during the 100-day summer season. Based on a Natural Resources Defense Council (NRDC) study, this was the worst ocean water quality problem in the nation. In 1989 a survey by the *New Jersey Star Ledger* newspaper ranked the shore water quality problem as the fourth most important in the state, after crime, drugs, and car insurance. In 1993, New Jersey had 90 beach-block-days of closures due to pollution—a reduction of the problem by almost 90 percent. This improvement in ocean water quality is the most significant in the nation based on a June 1993 NRDC study.

6. Method:

Root cause analysis found the problems were primarily from leaks and misconnections from a poorly maintained storm-drain and sewage infrastructure. The press had been focused on off-shore dumping and visible floatables as the causes of the problem.

The 90 New Jersey shore municipalities needed a measurable goal, support, and motivation to improve their storm drain and sewage infrastructure.

The QNJ team worked with the New Jersey DEP to develop a measurable goal. The measure was beach-block-days of closures due to pollution. The goal was zero by 1996. This measure and goal was based on a beachgoer survey and input. The goal gave us the direction we needed to work, and the annual measure provided an indicator of progress toward that goal. We had targeted achieving a 30 percent per year reduction in defects (beach-block-days of closures due to pollution), based on experience in applying our total quality approach at AT&T on business problems.

The team provided assistance on analyzing the ocean water quality data to determine root causes and prioritize them for improvement efforts based on the impact they had on the number of beach-block-days of closures caused.

The team provided assistance in obtaining best practice countermeasures from shore municipality experiences and sharing both the root causes and countermeasures with all the shore municipalities for replication.

The team created an annual QNJ Shore Quality Award to help communicate our goal, to motivate continuous improvement, and to learn and share best practices. The award was presented at an annual QNJ quality conference by the QNJ environment focus group chairperson and the commerce and/or DEP commissioner in New Jersey.

7. Results:

New Jersey ocean-water quality has achieved the most significant improvement in the nation, based on a 1993 NRDC study. A 90 percent reduction in our beach pollution problems in five years.

The citizens and media in New Jersey have been raving about the clean ocean water in New Jersey during the summer of 1993. (Press article was attached.)

The New Jersey shore tourism revenue has increased from $7 billion in 1989 to $11 billion in 1993, the fastest growing sector of the New Jersey economy.

New Jersey has a state sales tax of 6 percent. Consequently, the $4 billion increase in tourism revenue has resulted in an increase in state sales tax revenues of $240 million per year.

The QNJ team has recognized five shore town and one shore county for their improvement efforts.

Marine life is improved, and the dolphins are back in quantity.

This improvement has provided an example of how both New Jersey business and the New Jersey environment can be improved using a quality approach, which we hope to replicate in other areas with the support of the New Jersey DEP.

8. Metrics:

When the QNJ team started, the New Jersey DEP measured beach closures, which did not provide a good indication of either the length of the beach closed or the duration of the beach closure—both important measures to the beachgoer. Neither did it assess how significant our water quality problem really was. We worked with the New Jersey DEP, which agreed to change its measure to beach-block-days of closure, thereby providing a much better indication of the problem we were dealing with.

The team also obtained measures of shore tourism from the Commerce Department, as an indicator of the satisfaction of the beachgoers and the success of shore business as a result of annual improvements.

9. Benefits to customers:

Four customers have benefited:

- First, the New Jersey beachgoers are delighted with the New Jersey shore water quality improvement in 1993.
- Second, New Jersey shore business has benefited by an increase in tourism revenue of almost $1 billion per year—during a flat economy.
- Third, the state of New Jersey has benefited from an increase of $240 million per year in sales tax revenue as a result of the tourism revenue increase.
- Fourth, the marine life has benefited along the New Jersey shore. In 1988, it has been estimated that half the North Atlantic bottle-nosed dolphins (2,500 out of 5,000) died due to bacteria and virus in the polluted waters. In 1993, the dolphins are healthy and back in full force.

Individual Crusades

While writing this book I had the opportunity to read two wonderful books that I would highly recommend concerning efforts by individuals to improve the environment. One was *A Civil Action*, by Johnathan Harr, and the other was *And the Waters Turned to Blood*, by Rodney Barker.

I found myself trying to compare the stories in these other two books to the story that I had to tell.

A Civil Action is the story of a lawyer, Jan Schlichtmann, who devotes a significant part of his life to one case to determine who was responsible for the pollution of drinking water in Woburn, Massachusetts (only two towns away from where I grew up in Arlington, Massachusetts). The story is of an individual fighting to uncover the truth, to make those responsible pay those who were injured. In the end, based on evidence uncovered by Schlichtmann, the EPA finally agrees to undertake a cleanup action that was the most costly (an estimated $69.4 million) ever undertaken in New England.

And the Waters Turned to Blood is the story of Dr. JoAnn Buckholder, a scientist who discovers a new toxic organism that is responsible for a large part of the fish kills in North Carolina rivers and estuaries. She also determined that this toxic organism multiplies and grows in volume when the waters are polluted with agricultural, industrial, and municipal waste. She then faced enormous difficulties and delays in getting the system to take action on her findings.

Both of the above books are interesting human dramas about the struggle of an individual trying to uncover the truth and overcome the system to get someone to take action on the truth they have discovered. Both involve the individual bringing all their professional experience and knowledge to bear on the problem. In the process, we learn a lot about both professions—some good and some bad.

The Dolphins Are Back is not about an individual effort against the system, but about a team and partnership effort working with the system. It is not about one case, or one cause, of pollution; it is about many sources of pollution that all must be addressed. It is not about a single state or city department of environmental protection, or a few businesses that aren't responding quickly enough to fix pollution problems. It is about working with business and the federal, state, and county governments to achieve and sustain a continuous improvement effort at the local level. It is about using common-sense tools, called quality methods, and data to find the causes of problems and develop and deploy countermeasures as rapidly as possibly to prevent the recurrence of the problem.

Our lives and our economy are improved by preventing pollution, and not by "permitting" pollution. It is not about an individual professional effort, it is about encouraging many individual citizen efforts. Although the struggle of an individual against the system makes for a great story, citizens partnering with government to make the system work can produce even greater results. The government system we created requires citizen involvement to work. Our governance process also requires continuous improvement, by its citizens, to be capable of addressing today's and tomorrow's problems.

Endnotes

Chapter One

1. *New York Times*, September 27, 1987.
2. *The Economist*, November 28, 1987.
3. Scott, G. P. The dolphin die off: Long-term effects and recovery of the population. Proceedings, *Oceans*, 1988. Baltimore, MD, pp. 819–823.
4. Lipscomb, Thomas. "Morbilliviral Disease in Atlantic Bottlenose Dolphins from the 1987–1988 Epizootic." *Journal of Wildlife Diseases*, October, 1994, pp. 567–571.
5. Kannan, Kururntha, Dr. *Environmental Science & Technology*, January 1997.
6. Barker, Rodney, *And the Waters Turned to Blood*. New York: Simon & Schuster, 1997.
7. *Daily Record*, May 25, 1989.

Chapter Two

1. The Center for Marine Conservation (CMC), Washington, DC. Fowle, Susane, "Fish for the Future: A Citizens's Guide to Federal Marine Fisheries Management." 1993.

Chapter Three

1. Reprinted by permission from AT&T's PDCA Cycle (#500–799), copyright 1996. Published by AT&T Quality Office. For additional copies call 1-800-432-6600 or 1-317-322-6416.

Chapter Four

1. Improvement Act of 1987—Public Law 100–107.

Chapter Six

1. These designer plates can be ordered and received through the mail from the New Jersey DMV by calling 1-609-292-6500.

Chapter Seven

1. Baldrige Foundation directors from the education and health care sectors were added to the foundation in 1996. In 1997, fund-raising plans were developed by the Foundation to support the new award categories, with one new award to begin in 1998 and one to begin in 1999, provided Congress passed a bill in 1997 to provide its support for this business–government partnership.

   Unfortunately, on November 14, 1997, Congress did not approve the $2.2 million requested to expand the award to the education and health care categories. However, legislators directed $8.8 million to two state universities for research projects that bypassed the normal federal grant application process. This appears to be putting state special interests over national interests.

Chapter Eight

1. From Hewitt Associates survey of 1681 U.S. employers, 1997.
2. For current results, see National Quality Program Web site: http://www.qual1ty.nist.gov.
3. *New York Times*, March 1, 1997.
4. New Jersey DEP Office of Land and Water. "New Jersey 1992: State Water Quality Inventory Report," November 1993.
5. New Jersey DEP Bureau of Air Monitoring. "1993 Air Quality Report," May 1994.
6. New Jersey DEP. "Site Remediation Report—1994," 1995.
7. Kent E. Portney. *Controversial Issues in Environmental Policy.* Newbury Park, California: Sage Publications, Inc., 1992.
8. *The Sunday Star Ledger*, February 23, 1997.
9. *New York Times*, July 7, 1997.
10. *The Sunday Star Ledger*, February 23, 1997.

Chapter Eleven

1. In 1997, while at a quality conference in Washington, D.C., I was given a great little booklet on several successful applications of quality improvement in the federal government, *The Blair House Papers*, January 1997. This small booklet provides a short summary of many quality improvements made under the Reinventing Government program led by Vice President Al Gore. (*National Performance Review*, Telephone: 202-632-0150; Internet: www.npr.gov.)
2. *New York Times*, August 29, 1997.

Bibliography

Allen, Robert. Interview by author, 8 July 1997.

Ackerman, Rodney, Roberta Coleman, Elias Ledger, and John MacDorman. "Process Quality Management & Improvement Guidelines." AT&T Quality Office, 1988.

Annitto, Susan, and Dale Myers. "Plan–Do–Check–Act Cycle." AT&T Quality Office, 1996.

Barker, Rodney. *And the Waters Turned to Blood*. New York: Simon & Schuster, 1997.

Barton, Kimberly, and Dave Fuller. *Testing the Waters V (1994)*. Natural Resources Defense Council. New York: Desktop Publishing, 1995.

Barton, Kimberly, and Dave Fuller. *Testing the Waters VI (1995)*. Natural Resources Defense Council. New York: Desktop Publishing, 1996.

Barton, Kimberly, and Dave Fuller. *Testing the Waters VII (1996)*. Natural Resources Defense Council. New York: Desktop Publishing, 1996.

Commager, Henry, and Samuel Morison. *The Growth of the American Republic, Vol. One*. New York: Oxford University Press, 1962.

Coe, James, and Donald Rogers. *Marine Debris—Sources, Impacts, and Solutions*. New York: Springer-Verlag, 1997.

Conlin, Linda. "Travel and Tourism in New Jersey." New Jersey Division of Travel and Tourism, April 1997.

DiBiaggio, John. "President's Corner," *Boston Globe*, May 2, 1997.

Finamore, Nick. Interview by author, August, 1997.

Gavin, Thomas. "Characteristics of Federal Agency Operations which Affect the Performance Measurement Process," Chantilly, Virginia: National Reconnaissance Office, April 1997.

Geraci, Joseph D. "Clinical Investigation of the 1987–1988 Mass Mortality of Bottlenose Dolphins Along the U.S. Central and South

Atlantic Coast." Final report to National Marine Fisheries Service and Marine Mammal Commission, Ontario Veterinary College, April 1989.

Gore, Al. "The Blair House Papers." The National Performance Review, January 1997. (Foreword by Bill Clinton and introduction by Al Gore.)

Groves, Laurie. Interview by author, July 1997.

Hatala, Lew, and Marilyn Zuckerman. *Incredibly American*. Milwaukee, WI: ASQC Press, 1992.

Ianna, Frank. Interview by author, July 1997.

Jargowsky, Lester. Interview by author, July 1997.

Kannan, Kurumtha, Dr. *Environmental Science & Technology*, Michigan State University, January 1997.

Kirby, J. Philip, and David Hughes. *Thoughtware: Change the Thinking and the Organization Will Change Itself*. Portland, OR: Productivity Press, 1997.

Lewy, Guenter. *America in Vietnam*. New York: Oxford University Press, 1978.

Lipscomb, Thomas, Dr. "Morbilliviral Disease in Atlantic Bottlenose Dolphins from the 1987–1988 Epizootic." *Journal of Wildlife Diseases* (October 1994), pp. 567–571.

Olson, James. *The Dictionary of the Vietnam War*. Westport, CT: Greenwood Press, 1988.

Pagliughi, Martin. Interview by author, October, 1997.

Portney, Kent. *Controversial Issues in Environmental Policy*. Newbury Park, CA: Sage Publications, Inc., 1992.

Rayner, Wendy. Interview by author, August 1997.

Rosenblatt, Dave. "The Cooperative Coastal Monitoring Program," Annual Reports. Division of Water, New Jersey DEP, 1989 to 1996.

Rosenblatt, Dave. Interview by author, July 1997.

Sheavly, Seba. "Coastal Cleanup Results," Annual Reports. Center for Marine Conservation, 1989 to 1995.

Shinn, Bob. "Performance Partnerships for the Next Generation," New Jersey DEP 1996 Annual Report, 1997.

Soltis, Judy. Interview by author, July 1997.

Yaskin, Judith. "New Jersey Storm Water Act," New Jersey DEP, February 1988.

"New Jersey Sewage Infrastructure Improvement Act Phase II," New Jersey DEP, February 1988. (Prepared by Division of Water Resources.)

"Assessment of Floatables Action Plan," Region II EPA, May 1995. (Prepared by Water Management Division.)

"A Framework Document for a Coastal Management Partnership," New Jersey DEP Coastal Report Taskforce, 1997.

"Sustainable Growth Measures for New Jersey," New Jersey Future, 1997.

"1996 Environment Health and Safety Report," AT&T Environment Health and Safety Department, 1996–1997.

"*USA Today* RIT Quality Cup Application," Rochester Institute of Technology, 1993.

About the Author

Phillip Scanlan's career started with AT&T in 1966. Over the past 30 years he has served AT&T as a manufacturing engineer, a product planning manager, quality director, and vice president for quality. As a quality leader Mr. Scanlan has been instrumental in developing an AT&T Total Quality Approach supported by an AT&T Quality Library of 28 books and an AT&T Core Quality Curriculum of more than a dozen quality courses, which have been approved for college credits. Mr. Scanlan has also been responsible for the development of an annual AT&T Chairman's Quality Assessment Process to provide feedback that guides a continuous improvement effort. With Mr. Scanlan's assistance, AT&T is the only company to have more than two business units win the Malcolm Baldrige National Quality Award (two in 1992 and one in 1994).

Mr. Scanlan's interest in quality processes has led him to serve as a founding member and an officer of Quality New Jersey and chair of its Environmental Focus Group. In 1997, he was elected Chair for the Quality New Jersey organization. He has also been a leader of the AT&T–Rutgers University Quality Partnership.

Mr. Scanlan is a member of the American Society for Quality (ASQ) and was selected as the keynote speaker for the first joint ASQ/AMA conference. He was selected as a Malcolm Baldrige National Quality Award Examiner in 1990, served as a Senior Examiner in 1991, and was a member of the Baldrige Foundation's support team from 1995 through 1997. In 1993, Mr. Scanlan addressed the World Quality Congress in Helsinki, Finland. In 1997, Mr. Scanlan became a member of a Center for Marine Conservation (CMC) advisory board.

Mr. Scanlan has a bachelor of science degree in electrical engineering and a master of science degree in engineering management from Northeastern University. In 1988 he completed an Executive Program, "Managing the Enterprise," at Columbia University.

Serving in the U.S. Army Signal Corps from 1967–1968, he developed communications training courses in 1967 and ran a communications center in Nha Trang, Vietnam in 1968.

Index

BOOKS FROM PRODUCTIVITY PRESS

Productivity Press publishes books that empower individuals and companies to achieve excellence in quality, productivity, and the creative involvement of all employees. Through steadfast efforts to support the vision and strategy of continuous improvement, Productivity Press delivers today's leading-edge tools and techniques gathered directly from industry leaders around the world. Call toll-free (800) 394-6868 for our free catalog.

BECOMING LEAN
Inside Stories of U.S. Manufacturers
Jeffrey Liker

Most other books on lean management focus on technical methods and offer a picture of what a lean system should look like. Some provide snapshots of before and after. This is the first book to provide technical descriptions of successful solutions and performance improvements. The first book to include powerful first-hand accounts of the complete process of change, its impact on the entire organization, and the rewards and benefits of becoming lean. At the heart of this book you will find the stories of American manufacturers who have successfully implemented lean methods. Authors offer personalized accounts of their organization's lean transformation, including struggles and successes, frustrations and surprises. Now you have a unique opportunity to go inside their implementation process to see what worked, what didn't, and why. Many of these executives and managers who led the charge to becoming lean in their organizations tell their stories here for the first time!
ISBN 1-56327-173-7/ 350 pages / $35.00 / Order LEAN-B287

LEARNING ORGANIZATIONS
Developing Cultures for Tomorrow's Workplace
Sarita Chawla and John Renesch, Editors

The ability to learn faster than your competition may be the only sustainable competitive advantage! A learning organization is one where people continually expand their capacity to create results they truly desire, where new and expansive patterns of thinking are nurtured, where collective aspiration is set free, and where people are continually learning how to learn together. This compilation of 34 powerful essays, written by recognized experts worldwide, is rich in concept and theory as well as application and example. An inspiring follow-up to Peter Senge's groundbreaking bestseller *The Fifth Discipline*.
ISBN 1-56327-110-9 / 571 pages / $35.00 / Order LEARN-B287

BUILDING A SHARED VISION
A Leader's Guide to Aligning the Organization
C. Patrick Lewis

This exciting new book presents a step-by-step method for developing your organizational vision. It teaches how to build and maintain a shared vision directed from the top down, but encompassing the views of all the members and stakeholders, and understanding the competitive environment of the organization. Like *Corporate Diagnosis*, this book describes in detail one of the necessary first steps from *Implementing a Lean Management System*: visioning.
ISBN 1-56327-163-X / 150 pages / $45.00 / Order VISION-B287

BEYOND CORPORATE TRANSFORMATION
A Whole Systems Approach to Creating and Sustaining High Performance
Christopher W. Head

When do your employees resist change? They resist change when they don't understand the changes that are taking place, they see little or no perceived benefit of doing things differently, or they do not feel involved. Which is why employees who will be affected by a transformation must effect the changes. Realizing that anything short of total employee involvement in the change process jeopardizes success, this book emphasizes that it is the responsibility of every employee to act as a change agent. Learn how to go beyond piecemeal incremental changes, beyond reengineering, beyond the limited idea of change to encompass a whole systems approach to creating and sustaining a competitive advantage. Through a revolutionary, integrated, employee-oriented leadership philosophy, this book illustrates how to transform an organization by tapping into the full potential of every employee.
ISBN 1-56327-176-1 / 240 pages / $35.00 / Order BEYOND-B287

THOUGHTWARE
Change the Thinking and the Organization Will Change Itself
J. Philip Kirby & D.H. Hughes

In order to facilitate true change in an organization, its thinking patterns need to be the first thing to change. Your employees need more than empowerment. They need to move from doing their jobs to doing whatever is needed for the good of the entire organization. Thoughtware is the underlying platform on which every organization operates, the set of assumptions upon which the organization is structured. When you understand and change thoughtware, the tools and techniques of continuous improvement become incredibly powerful.
ISBN 1-56327-106-0 / 200 pages / $35.00 / Order THOUG-B287

TOOL NAVIGATOR
A Master Guide for Teams
Walter J. Michalski

Are you constantly searching for just the right tool to help your team efforts? Do you find yourself not sure which to use next? Here's the largest tool compendium of facilitation and problem-solving tools you'll find. Each tool is presented in a two- to three-page spread which describes the tool, its use, how to implement it, and an example. Charts provide a matrix to help you choose the right tool for your needs. Plus, you can combine tools to help your team navigate through any problem-solving or improvement process. Use these tools for all seasons: team building, idea generating, data collecting, analyzing/trending, evaluating/selecting, decision making, planning/presenting, and more!

ISBN 1-56327-178-8 / 550 pages / $150.00 / Order NAVI1-B287

TO ORDER: Write, phone, or fax Productivity Press, Dept. BK, P.O. Box 13390, Portland, OR 97213-0390, phone 1-800-394-6868, fax 1-800-394-6286.

Outside the U.S. phone (503) 235-0600; fax (503) 235-0909

Send check or charge to your credit card (American Express, Visa, MasterCard accepted).

U.S. ORDERS: Add $5 shipping for first book, $2 each additional for UPS surface delivery. Add $5 for each AV program containing 1 or 2 tapes; add $12 for each AV program containing 3 or more tapes. We offer attractive quantity discounts for bulk purchases of individual titles; call for more information.

ORDER BY E-MAIL: Order 24 hours a day from anywhere in the world. Use either address:

To order: **service@ppress.com**

To view the online catalog and/or order: **http://www.ppress.com/**

QUANTITY DISCOUNTS: For information on quantity discounts, please contact our sales department.

INTERNATIONAL ORDERS: Write, phone, or fax for quote and indicate shipping method desired. For international callers, telephone number is 503-235-0600 and fax number is 503-235-0909. Prepayment in U.S. dollars must accompany your order (checks must be drawn on U.S. banks). When quote is returned with payment, your order will be shipped promptly by the method requested.

NOTE: Prices are in U.S. dollars and are subject to change without notice.